Milk and Milk Products Technology

Milk and Milk Products Technology

Subhasish Biswas MVSc, PhD
Professor and Head
Department of Animal Products Technology and Marketing
Faculty of Veterinary and Animal Sciences
West Bengal University of Animal and Fishery Sciences
37, KB Sarani, Belgachia
Kolkata 700 037

Subhash Kumar Bhattacharyya MSc, PhD, AIC
Former Deputy Milk Commissioner
Govt. of West Bengal and Visiting Lecturer
Department of Animal Products Technology and Marketing
West Bengal University of Animal and Fishery Sciences
37, KB Sarani, Belgachia
Kolkata 700 037

JAYPEE BROTHERS MEDICAL PUBLISHERS (P) LTD

New Delhi • Panama City • London

Published by

Jaypee Brothers Medical Publishers (P) Ltd

Corporate Office

4838/24, Ansari Road, Daryaganj, **New Delhi** 110 002, India
Phone: +91-11-43574357, Fax: +91-11-43574314
www.jaypeebrothers.com

Offices in India

- **Ahmedabad**, e-mail: ahmedabad@jaypeebrothers.com
- **Bengaluru**, e-mail: bangalore@jaypeebrothers.com
- **Chennai**, e-mail: chennai@jaypeebrothers.com
- **Delhi**, e-mail: jaypee@jaypeebrothers.com
- **Hyderabad**, e-mail: hyderabad@jaypeebrothers.com
- **Kochi**, e-mail: kochi@jaypeebrothers.com
- **Kolkata**, e-mail: kolkata@jaypeebrothers.com
- **Lucknow**, e-mail: lucknow@jaypeebrothers.com
- **Mumbai**, e-mail: mumbai@jaypeebrothers.com
- **Nagpur**, e-mail: nagpur@jaypeebrothers.com

Overseas Offices

- **Central America Office, Panama City, Panama**, Ph: 001-507-317-0160
  e-mail: cservice@jphmedical.com, Website: www.jphmedical.com
- **Europe Office, UK**, Ph: +44 (0) 2031708910
  e-mail: info@jpmedpub.com

Milk and Milk Products Technology

First Edition: **2006,** Reprint: 2023

ISBN 81-8061-877-3

Typeset at JPBMP typesetting unit

Printed in India

To
the memory
of
our beloved parents

West Bengal University of Animal and Fishery Sciences

68, Kshudiram Bose Sarani, Kolkata- 700 037, Phone : 2556 3450, Resi : 2680 1754,
Gram : VETUNIV, Fax : 91-033-2557 1986

Dr. A. K. Bandyopadhyay
Vice-Chancellor

No : VCS/WBUAFS/P/26/114
Date : 12.5.06

Foreword

The book entitled 'Milk and Milk Products Technology' is an excellent compilation of different areas of milk and its sciences. The fiften chapters vividly elaborated in this book will definitely fulfill the requirements of undergraduate and postgraduate students studying in the discipline of Livestock Products Technology under the curriculum of Veterinary Council of India.

India has made a significance progress in fluid milk production and presently occupying as the first position in the world, which is a reflection of blessed touch of science and technology in this particular area of livestock production. Still many more things are to be carried out to make this country not only to progress in livestock production but to alleviate the poverty of the rural poor through this branch of science and technology. The technology of milk products will definitely be an area of entrepreneurship for an economical and industrial growth of our country.

It is pleasing to note that Dr Subhasish Biswas, Professor and Head, Deptt. of Animal Products technology and Marketing, Faculty of Veterinary and Animal Sciences, West Bengal University of Animal and Fishery Sciences, Kolkata, and Dr Subhash Kumar Bhattacharyya, Former Deputy Milk Commissioner, Govt. of West Bengal and now visiting lecturer in the Deptt. of Animal Products Technology and Marketing, Faculty of Veterinary and Animal Sciences, West Bengal University of Animal and Fishery Sciences, Kolkata, the excellent workers in the branch of science have made such compact compilation of work encompassing the need of the students, both PG and UG in this book of 'Milk and Milk Products Technology' which finds expression in the fifteen chapters of this book.

I wish and acknowledge this benevolent endeavour of the authors a grand success.

(Prof.) AK Bandyopadhyay

SHER-E-KASHMIR
UNIVERSITY OF AGRICULTURAL
SCIENCES AND TECHNOLOGY OF JAMMU
Rail Head Complex, Jammu-180 012

Phone: (O) +91-191-2473883, 2471745, 2473417
Telefax: +91-191-2473883
Resi.: +91-191-2434882, 2457401
E.mail: vc_nsharma@rediffmail.com
Website: www.skuast.org

Dr. Nagendra Sharma
MVSc, PhD, DSc, FNAVS, FIAAVR
VICE-CHANCELLOR

D.O. No.:

Dated: 12-5-20

Foreword

India is bestowed with the blessing of 'Operation Flood' the world's biggest dairy development programme in terms of coverage and longevity. It has been a powerful development miracle by creating a stable outlet of milk produced in rural areas, providing additional income to villagers, Improved livelihood security and living standards and status to India as the World's highest milk producing country.

In this scenario of Science and Technological upliftment in this particular area, it is a great pleasure that a book entitled "Milk and Milk Products Technology" is going to be published encompassing a number of aspects of milk and milk products within its domain, in fifteen chapters. The total writings have been done in a lucid manner.

This will be of a considerable help and guidance to the students, teachers, entrepreneurs as well as the technologists related to the 'White blessing' of mother earth.

I appreciate the endeavour of Prof S Biswas and Dr SK Bhattacharyya of West Bengal University of Animal and Fishery Sciences, Kolkata- 37 for preparing this compilation with immense care and through introspection.

I consider this is a unique opportunity to introduce this book to the benefit of all the students and persons dealing with milk—its science and technology.

I wish the venture a great success.

Nagendra Sharma

Preface

The role of 'Milk and Milk Products' has always been enjoying mass appeal. Traditional Dairy Products have played a significant role in the economic, social, religious and nutritional well-being of our people since time immemorial.

Realising that transfer of available technologies to the field being a very low ebb, we have tried to conceive the same while publishing this small but informative handbook of fifteen chapters, catering the requirements of undergraduate and postgraduate students studying in the discipline of Livestock Products Technology under the Curriculum of Veterinary Council of India.

In preparing this handout, the authors aimed to cover both the theoretical and the practical phases of some areas of market milk production and the processing of related products which are of frequent use in the dairy industry. The "Milk and Milk Products" handout attempts to correlate the applications of science with milk-plant operations. This handout has been written in clear and coincise manner, supplemented with tables, photographs of some of the equipment commonly used in the dairy industry, having enriched with updated informations.

The first draft of this manuscript was prepared a year back. Since then the process of accumulating information in the related field and the effort of knitting them together continued. All these resulted in its present form. Here, the helps of PG and PhD Scholars of our Department to orient the whole thing are gratefully acknowledged. Without their sincere help in compilation and typing, it would not have been possible to make this venture a success. Constant moral support for preparing the manuscript from Dr. A Goswami, our friend, is acknowledged with gratitude. The diagrams drawn by Dr. A Chakraborty, one of the PhD scholars are self explanatory for admiration.

The critical comments and any suggestion from the users on the materials presented will go a long way in improving the future editions of this book. Last but not the least; Prof. S Biswas is indebted to his wife 'Anuradha' and daughters 'Olipriya' and 'Arpita' for their endurance to facilitate such compilation. Dr Bhattacharyya also expresses his sincere gratitude to his wife 'Subhra' and daughter 'Sarmistha' for their supports in preparing the manuscripts.

Authors are thankful to M/S Jaypee Brothers, New Delhi for their innate interest of disseminating the information of this field by publishing this book.

Subhasish Biswas
Subhash Kumar Bhattacharyya

Contents

CHAPTER 1

Development of Dairy Industry in India

India has been, for thousand of years, a dairying country attaching considerable importance to production and consumption of milk. The conversion of milk into products such as dahi, butter, and ghee was well known even in vedic times. Boiling of milk for prolonging its life and making it safe for human consumption was commonly practiced.

Before the first world war of 1914, there was practically no existence of dairy industry in the country. As a result of the demand for some western type products during the war, a few military dairy farms were started in different parts of the country for meeting the needs of military personnel. Since then indigenous cows were poor milk yielder and buffalo milk was a strange commodity to the British soldiers, cross-breeding of local cows with British breeds of dairy cattle was started along with maintaining small herds of pure exotic animals in high altitude areas.

The private industry soon saw the bright opportunities in the dairy business and entered the area in 1929. Some pioneer organisations like Polsons Ltd., Edward keventers are worth to mention and they catered mainly the requirements of defence establishments and to a small section of the elite civilian population. The Government of India also set up a rural milk condensary at Anand as a subsidiary of Imperial Dairy Research Institute. Milk was being supplied all over the country primarily as raw milk at the consumer's door step in the traditional way till late 1950's.

Organised dairy development started in our country only after independence and was attained in 1947. The first large-scale distribution of processed and bottled milk was in Supply Greater Bombay Milk Scheme in 1949. The Kaira district co-op. milk producer's union (AMUL) took the responsibility for procurement, processing and despatch of milk from Anand to Bombay to supplement the milk locally produced. A product manufacturing plant to handle surplus milk was started at Anand with the aid of UNICEF in 1954.

The dairy development wing of the ministry of agriculture and the Planning Commission recognised the need of developing dairy industry and started to make specific financial outlays beginning with the First Five Year Plan (1956-1961) for starting dairy schemes, setting up city milk dairies, and milk collection- cum-chilling centres. Cattle development project of animal husbandry wing of ministry of agriculture and state animal husbandry department lent support to milk production programmes.

Realising the increasing gap between the needs and availability of milk for the city dwellers, and the need for supplying milk of hygienic and reliable quality at a reasonable price, the Government started a number of 'milk supply schemes' for cities with population of 5,00,000 and above initially, then extended these to cities with 1,00,000 and above under state Five-year Plan allocations. The allocations for dairying under the Ministry of agriculture was thus increased to Rs.190 million in the 2nd Plan, Rs. 360.5 million in the 3rd, and rose gradually to Rs. 4,623 million in the 6th Plan.

To overcome some of the constraints Government then set up "State Dairy" Corporations, which were, bestowed a greater degree of operational autonomy but under the control of the Government. In order to develop the dairy industry in the country, the National Dairy Development Board (NDDB) was set up towards the end of 1964, by Government of India as an autonomous body of professional experts. Since by its constitution, the NDDB is a non-profit making body public trust, the Govt. of India set up the Indian Dairy Corporation (IDC) a wholly Govt. owned body and the NDDB along with IDC launched the Operation Flood Programme-I in 1970 and later on the Operation Flood II in 1979 with financial outlays of Rs. 1160 million and Rs. 4850 million respectively.

Dairy in India has meant our people, since ancient days, milk production primarily through cow keeping and using the milk for the use of the family and for the religious rituals. The surplus quantities of milk were conserved by converting the milk into dahi, which after accumulation was churned in to butter, utilising butter milk for family consumption and converting surplus butter into ghee which was used as a part of food or for rituals.

Dairy development has now become a household often used word in various rural development programmes. The programmes and projects for development of dairy industry in our country aimed at organised milk supply to urban towns are about 40 years old. Prior to 1950, almost entire milk supply to consumers in urban areas was a monopoly in the hands of the unorganised sectors particularly the middlemen. There was no consideration of quality, either chemical or bacteriological.

From this stage, the country has marched forward in the short span of 30 years and there were about 250 organised dairies supplying milk to urban consumers (1986). Albeit, only 15 per cent of the total milk production was estimated to be handled by the organised dairy sector has captured 50 to 75 per cent of the liquid milk market share of the metropolis viz. Bombay, Calcutta, Madras, and Delhi.

In Eastern India including the north-eastern part, production of milk is not adequate to meet the demand. However, over the last 20 years, the situation has spectacularly changed in most of the states. States like Rajasthan, Madhya Pradesh, Andhra Pradesh, Karnataka, Tamil Nadu, and even Kerala has generated marketable surpluses. The states in eastern part of India, however, lack behind. The factors in different parts of India, which impede rapid dairy development are:

- Agro-climatic factors
- Communication problems
- Economic factors
- Sociological factors
- Inadequate infrastructure
- Interference by powerful vested interests
- Lack of local leadership and professionals
- Lack of work ethics amongst employees
- Lack of remunerative market.

Linking the rural producer with the urban consumer is one of the major objectives of rural modernisation. With the development of transportation and communication systems, reaching the major markets has become a reality and has opened opportunities for the producers to obtain better price for their produce. From the Anand-Bombay milk line, which established the first direct link between the consumers of Bombay and milk producers, in the hinterland milk shed of Kaira, one of the most significant achievements of the 1950s, has emerged as the cross-country National Milk Grid (NMG) system. No doubt, the per capita milk availability, which dropped from 132 g in 1951 to 107 g in 1969-1970, has increased to 163 g (1990). However, the regional and seasonal imbalances in our milk production do not permit this level of availability across the country. The sensitivity of the lean and flush characteristics of milk supply, with demand remaining constant year round, makes the development of dairy industry a real challenge and an opportunity.

Milk Production Scenario in India

India has emerged as the world's largest milk producing country. India's milk production during 2001-2002 could rise to an estimated annual figure of 81 million tonnes from the 79 million tonnes and 77 million tonnes of the previous two years respectively. It is projected that during 2009-2010 the milk production in our country is expected to reach 115 million tonnes at an expected growth rate of 4 per cent in milk production.

Per capita Availability of Milk for Consumption

With the growth of the human population in India at a compounded rate of 1.86 per cent and the annual growth rate in milk production at 4 per cent, the country is expected to attain self-sufficiency by 2007-2008 in providing 250 g of milk to its population.

Table 1.1: Milk production in India and per capita availability of milk

	1951	1971	1981	1999	2000-2001	2001-2002	2002-2003	2004
Annual milk production at growth rate (million tonnes)	17.4	22.53	34	77.69	80.80	84.03	87.39	90.88
Per capita availability of milk per day (in g)	132	114	128	214.32	218.82	223.42	228.11	232.90

Table 1.2: Consumption pattern of milk and milk product

Product	*Consumption items of per centage*
Liquid milk	46
Ghee	28
Curd	7
Butter	6.58
Khoa	5.5
Milk powder	3.8
Cheese	2.0
Others	1.12

Table 1.3: Value of output in dairy industry

Year	*Value (Rs. million)*
1951	8700
1961	10200
1971	19125
1981	75460
1985	126900
1986	131700
1990	204000
2000	325000

Table 1.4: Outlay and expenditure on dairying in various plan-periods (Rs. million)

Plan period	*Outlay*	*Expenditure*
1st plan (1951-52 to1955-56)	78.1	77.8
2nd plan (1956-57 to1960-61)	174.4	120.5
3rd plan (1961-62 to1965-66)	360.8	336.0
4th plan (1966-67 to1968-69)	261.4	257.0
4th plan (1969-70 to 1973-74)	1390.0	787.5
5th plan (1974-75 to 1977-78)	-	540.5
6th plan (1980-81 to 1984-85)	4603.0	4362.0
7th plan (1985-86 to 1989-90)	4934.7	4302.9

Annual report, Ministry of Agriculture GOI

Table 1.5: Milk production: World vs India (million tonnes)

Year	*World*	*India*	*Per cent to world total*
1980-81	457.7	31.6	6.9
1984-85	454.1	41.5	8.4
1985-86	504	44	8.7
1986-87	512	45.6	8.9

(FAO production year book)

Table 1.6: Milk production in India

Year	*Million tonnes*
1950-51	17
1960-61	20
1980-81	31.6
1985-86	44
1986-87	45
1987-88	46
1988-89	48.7
1989-90	51
1990-91	53.89
1991-92	56.45
1992-93	59.13
1993-94	61.94
1994-95	63.20
1995-96	66.05
1996-97	70
1997-98	74
1998-99	75.60
1999-2000	77.69
2000-01	80.80
2001-02	84.03
2002-03	87.39
2003-04	90.88

Table 1.7: Per capita milk availability g/day

1951	132
1956	135
1961	127
1966	108
1969-70	107
1971-72	112
1977-78	123
1978-79	124
1979-80	126
1980-81	128
1981-82	131
1983-84	138
1984-85	144
1989-90	165
1990-91	173
1999-2000	214.32
2000-01	218.82
2001-02	223.42
2002-03	228.11
2003-04	232.90

Table 1.8: Region-wise milk production and availability (1986-87)

Region	*Population in million*	*Milk production in million tonnes*	*Per capita availability g/day*	*Per cent of total*
Northern	217	19.53	247	44.5
Western	108.8	5.73	144	13.1
Southern	182.5	9.22	188	21
Central and eastern	258.1	9.42	100	21.4

(FAO year book-1986)
(IDM-1990)

Table 1.9: Livestock resources in India (FAO, 1994)

Species	*Population (million)*	*Percent of world population*	*World ranking*	*Contribution to total milk production in India (per cent)*
Cattle	192.9	15	1	33.6
Buffalo	78.8	50	1	63.6
Sheep	44.8	-	6	-
Goat	118.3	15	1	2.8

OPERATION FLOOD

Operation flood, the words of the United Nations report on its first phase, "is the largest dairy development project in the world". It is also, so the report says, the project "to which the World Food Programme made its biggest allocation of aid"

The name was conjured upto describe the aim – to create a flood of milk from many small farmer producers in the rural country side of India to form a new supply for the growing towns and the great cities, creating a new source of wealth for the villages and a better milk supply for the towns.

The replication of the Anand pattern of dairy co-operative in India through the use of food aid initially supplied by the World Food Programme and later by the EEC (European Economic Commodity) combined with a World Bank loan is known as "Operation Flood". The IDC (Indian dairy corporation) was designated as the authority responsible for the implementation of "Operation Flood".

Object

To provide rural milk producers a direct access to the urban markets so that the largest possible share of consumer's rupee spent on the milk and milk products could be transferred back to the rural milk producers.

Operation Flood (OF)

The key task was to establish replicability of Anand pattern dairy co-operatives structure outside Gujarat state.

Operation Flood Phase I

Formally launched on July 1,1970, involving an investment of Rs. 95.4 crores. It was the largest dairy development programme launched anywhere in the world.

The project was originally formulated for five years but it suffered delays and was therefore, extended. It completed on 31st March, 1981. The revised final allocation was Rs. 116.4 crores.

The object of this project was basically to streamline the milk supply to the public sector dairies in the four main cities of Bombay, Calcutta, Delhi, and Madras. The modern dairies in four cities increased their throughput from 9.02-lakh lit/day in 1970 to 23.43 lakh lit/day in March, 1981.

Operation Flood Phase II

Operation flood II launched on October 2,1979 was designed to create, on the foundation built by phase I, a viable dairy industry to serve nations need for milk and milk products during 1990s .

The programme of work was set out in two parts: July, 1978 to September 1979 to carry out pre programme action Rs. 485.5 crores approved by planning commission, and GOI with revised target and out lay of Rs. 273 crores.

OP-II was implemented and completed during the sixth five-year plan and completed on 31st March.

Operation Flood Phase III

Implemented during the 7th Five-year plan period, from April 1985 to March 1990.

The objects:

1. To intensify each union's work so as to bring more producers under the ambit of dairy co-operative
2. To ensure development of financially viable federations and unions
3. To intensify urban milk marketing
4. To provide technical inputs and infrastructure to union.

The outlay of the order of Rs. 681.29 crores which includes a provision of Rs. 100.0 crores to be revolved as working capital.

Overall Achievement through Operation Flood Programme

a. Milk production: Indians milk production having remained static around 20 million tonnes between 1951 and 1970, gone upto 51 million tonnes in 1990.
b. Per capita availability: As a result of increased milk production, which has been going up at 6% per annum, per capita consumption of milk which has come down to 132 g from 1950 to 107 g in 1970, has gone up to 174 g in 1990.
c. Milk shed coverage: In 1970, there were 5 milk shed cooperatives and under operation flood the number of milk shed has gone upto 174 in 1990.
d. Village level milk co-operatives: There were less than 2000 village co-operatives at the beginning of operation flood in 1970, which has gone upto 61,000 in 1990.
e. Rural milk procurement: The average procurement of milk by the dairy co-operatives in India was only half a million litre per day in 1970, which has gone upto 9.8 million litre per day in 1990.
f. Milk marketing: The marketing of milk as fluid milk by dairy co-operatives in nearly 200 cities and towns in India has gone up from merely a million lit/day in 1970 to 7.3 million lit/day in 1990.

g. Milk powder production and import: Milk powder production in India which was around 22000 tonnes in 1970 has gone upto 165000 tonnes. The import of milk powder which was around 60,000 tonnes in 1960 have now eliminated.

Table 1.10: Operation flood at a glance

Launched	(1970-81) July 1970	(1978-85) Oct 1979	(1985-1990) April 1985
Outlay	Rs. 95.4 Crores	Rs. 485.5 Crores	Rs. 681.29 Crores
Completed	31st March 1981	31st March 1985	31st March 1990

Table 1.11: Achievements through operation flood programme

	1970	1990
Milk production	20 million tonnes	51 million tonnes
Per capita availability	107 gm/day	173 gm/day
milk shed coverage	5 milk shed co-operative	174 milk shed co-operative
Village milk co-operatives	2000	61000
Rural milk procurement (by co-operative)	0.5 million lit/day	9.8 million lit/day
Milk marketing (by co-operative)	1.0 million lit/day	7.3 million lit/day
Milk powder production	22000 tonnes	165,000 tonnes

What is Anand Pattern Dairy Co-operative?

Dairying as a whole consists of four very important interrelated elements; milk production, procurement, processing, and marketing. The integrated approach to dairying including the above elements was adopted by the Kaira District Co-operative Milk Producers Union limited, Anand, popularly known as 'Amul' (Anand Milk Union Ltd.), which was set up in 1946. Both village and district dairy co-operatives were owned and controlled by the milk producer's themselves. This integrated approach to co-operative for dairying proved to be a great success with 'Amul', and naturally, was adopted by many other districts of Gujarat. This approach to dairying later came to be known as Anand pattern of dairy co-operative.

The essence of the Anand pattern is that the milk producers own and run the dairy organisations, performing all the functions of dairying. This, however, did not find favour with vested interests.

ANAND PATTERN MILK CO-OPERATIVE

STRUCTURE

Primary milk producers co-operative societies (MPCS) at the village level

↓

District co-operative milk producers union (DCMPU) at the district level

↓

Co-operative milk producers federation at the state level

SCHEMATIC REPRESENTATION

Producer

↓

Society

↓

Union

↓

Federation

↓

Consumer

ANAND PATTERN MILK CO-OPERATIVE PRINCIPLES

Anand pattern co-operatives are based on six co-operative principles:

1. Voluntary membership
2. Democratic decision-making

3. Limited interest on share capital
4. Equitable distribution of surplus
5. Co-operative education
6. Mutual co-operation.

Thus their functional dimensions encompass not only the economic but the social and moral obligations towards their members. Given these sets of conditions, it follows that co-operatives are an example of a perfect democratic institution of the members, for the members and by the members and their decision-making is based on one-member one-vote system.

A Typical Anand Pattern Co-operative Structure

A typical Anand pattern dairy co-operative structure has three tiers of well structured organisations with a milk producer constituting the smallest unit of the entire business enterprise. The three tiers are:

1. The village level dairy co-operative society federating producer members.
2. The district level co-operative milk producer's union federating the village societies and
3. The state level federation constituting of all the district level unions.

Of the tiers, the district co-operative union is the most active unit because it owns the physical infrastructure required for milk procurement, processing, manufacturing milk products, and for generating inputs required for increasing milk production.

The union integrates milk production, procurement, processing, and undertakes marketing either independently or with the help of apex federation.

West Bengal Women's Dairy Co-operative Project

Main Features at a Glance

1. Area of operation: 6 districts covered 4 milk unions.
   - Darjeeling
   - Jalpaiguri
   - Murshidabad
   - Nadia
   - 24 Parganas (North)
   - Burdwan
2. Project object: A project to empower rural women through the medium of dairy co-operative societies.
3. Launched: 1st April 1993.
4. Project duration: 3 years.
   1st year-April 1993 to March'94.
   2nd year-April'94 to March'95.
   3rd year-April'95 to March'96.
5. Estimated cost: 4 Crores 53 Lakhs 46 Thousand
6. Scheme: 90 per cent loan from central govt. and 10 per cent from the respective milk unions.
7. Target:
   1. 15000 women members in 125 units.
   2. 18000 kg milk/day (average milk collection).
8. District-wise WBDCP:

Dist.	Milk union
a. Himul-(40)	Darjeeling
b. Bimul (40)	Murshidabad
c. Kishan (35)	Nadia & 24 parganas(N)
d. Burdwan (10)	Burdwan
(Milk union 125)	

9. Technical set-up:
   i. One lady manager for each union
   ii. One lady project director at federation office

10. Organisations involved:
    i. Ministry of human resource development department of women and child development.
    ii. National dairy development board.
    iii. NABARD.
    iv. Department of animal resources, Govt. of West Bengal.
    v. Department of rural development, Govt. of West Bengal.
    vi. Department of health and primary welfare, Govt. of West Bengal.
    vii. Department of non-conventional energy and department of science and technology, Govt. of West Bengal.
    viii. Concerned district co-operative milk produces unions.
    ix. Milk commissioner Govt. of West Bengal.

The technology mission for dairy development. National Dairy Development Board and the National Milk Grid to be included in this chapter.

National Dairy Development Board (NDDB)

i. Registered in Sept 1965, under the Societies Registration Act, the Charitable Act and the Public Trust Act.
ii. Set up in 1965 to replicate the Anand Pattern Dairy Co-operations.
iii. Set up with its head quarters at Anand under the aegis of the Ministry of Agriculture, Govt. of India.
iv. Set up to make available multidisciplinary, professional expertise in the Public and Co-operative Sectors.
v. Assisted financially by the Govt. of India, Royal Danish Govt. and the Kaira Dist. Co-operative Milk Producer's Union Ltd. at the initial stage.
vi. It received aid from UNICEF in the form of teaching aid and equipments.
vii. Gujarat Govt. provided 25 acres of land for construction of NDDB office.

viii. Autonomous body-run by the decision of the Board of Directors.

ix. The head quarter is at Anand.

Structure and Organisation

A i. Policy making body is the Board.

ii. Chairman of the Board is appointed by the President of India.

iii. Members of the board are also appointed by the President of India in consultation with the chairman.

iv. The board can consist of a max. of nine members.

v. The board is assisted by an executive committee of 5-6 members including the chairman and the secretary of the board.

B. NDDB has established - 4 regional offices at Bombay, Bangalore, Calcutta and Delhi for implementing projects.

C. NDDB has 4 cell offices at Bhopal, Hydrabad, Madras and Patna for implementing projects in their areas of operation.

D. NDDB manages several units through Management Committee constituted by the executive committee, e.g.

a. Mother Dairy Calcutta

b. Mother Dairy Delhi

c. Patna Dairy Projects

d. Indian Immunologicals

e. Bhavnagar, Vegetable Products Unit

f. Sugam Dairy

g. Hindustan Packaging Company

h. Sabarmati Ashram Gaushala

i. Oilseeds and Vegetable Oil Wing

j. Indian Dairy Machins Company.

E. The head office of NDDB-Anand is organised into nine functional divisions/units:

1. Applied Research and Development
2. Computer Services

The Organisation

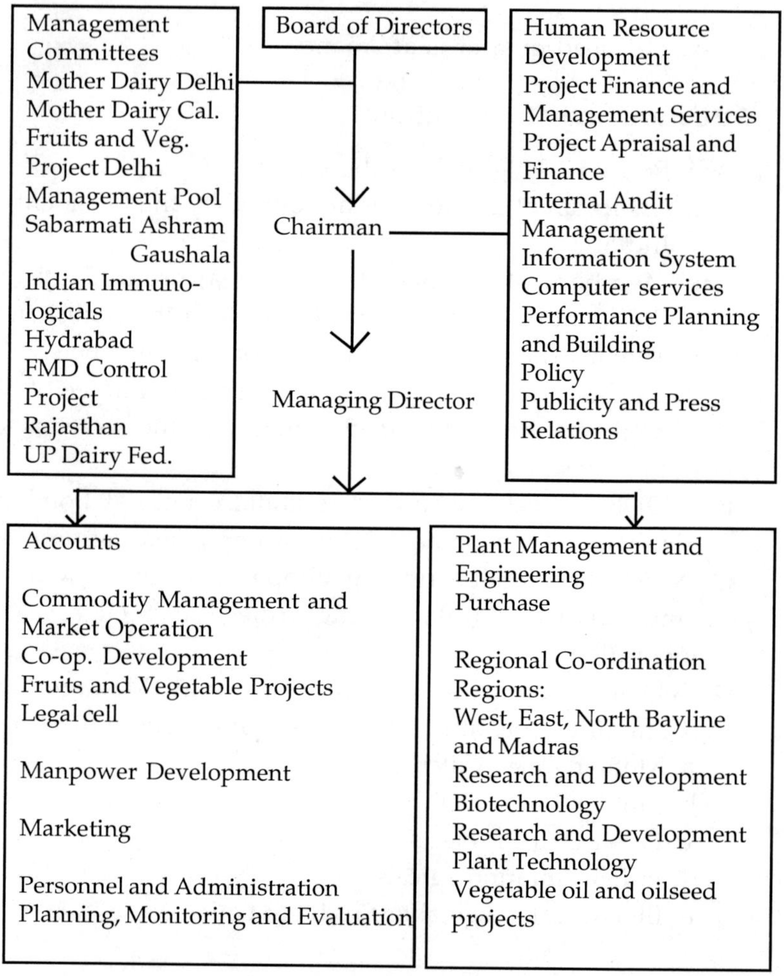

3. Engineers
4. Farmer's Organisation
5. Management and Economics Unit
6. Manpower Development
7. National Milch Herd
8. Planning cell
9. Purchase cell.

Objectives/Functions of NDDB

1. To sponsor, promote, install any plant/projects relating to dairying, animal husbandry, food and agriculture, fisheries and cold stores.
2. To undertake research work and development related to production and marketing of agriculture and food products.
3. To make available, on request, information, skills and technical services needed to increase milk production.
4. To make available technical expertise for boosting up procurement, processing and distribution activities.
5. To prepare initial feasibility studies of dairy and other agriculture related projects.
6. To provide manpower development services for dairy projects.
7. To provide technical programmes for training of personnel.
8. To help in the selection of equipment and to undertake its bulk procurement.
9. To offer consultation services on dairy and allied operations.
10. To provide international liason with other National Boards, agencies.

The Technology Mission for Dairy Development

Realising the importance of dairying in providing income and imployment to the rural poor, the Govt. of India has constituted a 'Technology Mission' for Dairy development. The Mission has set for itself certain key targets on milk production and marketing to be achieved by 1995. One of the major objectives of the Mission is to improve the availability of milk to the urban consumer and to provide stability to supplies by the application of appropriate, technologies in a variety of areas, establishment of large, energy efficient and cost effective dairies and development of technologies for the production of a range of indigenous dairy products.

Technology Mission for Dairy Development: Marketing

Targets	*Targets: 1995*
Components	
Milk procurement:	
Average - (MLPD)	15.0
Peak - (MLPD)	20.0
Milk marketing:	
Liquid milk - (MLPD)	11.3
Conserved commodities (LME)	2.7
Other products (MLPD)	1.0
Processing capacity (MLPD)	22.0

MLPP: Million lit/day

LME: Liquid milk equivalent in MLPD.

The National Milk Grid (NMG)

One of the most significant achievements of the 1950s, was to establish a direct link between the consumers of Bombay and the milk producers, in the hinterland milkshed of Kaira-emerging as a system - Known as cross - country 'National Milk Grid'.

During the first phases the NGM was to facilitate smooth supplies of liquid milk initially to the four Metro-cities and subsequently, to cover other important marketing centers.

CHAPTER 2

Composition and Nutritive Value of Milk

The Webster's dictionary defines milk as "a white or yellowish liquid consisting of small fat globules suspended in a solution, secreted by the mammary glands for the nutrition of the newborn".

The US Public Health Service (1967) defines milk as "milk is the lacteal secretion practically free from colostrums, obtained by the complete milking of one or more healthy cows, which contains not less than 8 ¼ per cent of milk solids not fat and not less than 3¼ per cent milk fat."

Milk is the diet whose sole function in nature is food. The role of milk is to provide nourishment and protection for the mammalian young. Milk also has been a food source for humans being since the dawn of history. Milk is a very complex food with over 100,000 molecular components.

Milk is an emulsion of fat globules and a suspension of casein micelles (casein, calcium, phosphorus), all suspended in an aqueous phase, which contains solubilised lactose, whey proteins, and some minerals, leucocytes in milk are part of the suspended phase.

The composition of milk of all mammals consists of the same constituents, though in varying proportions. The milk is composed of water, carbohydrate, fat, protein, minerals, and vitamins. It is important to remember that milk is secreted as a complex mixture of these components. The properties and importance of milk are greater and more complex than the sum of its components.

Composition of Milk

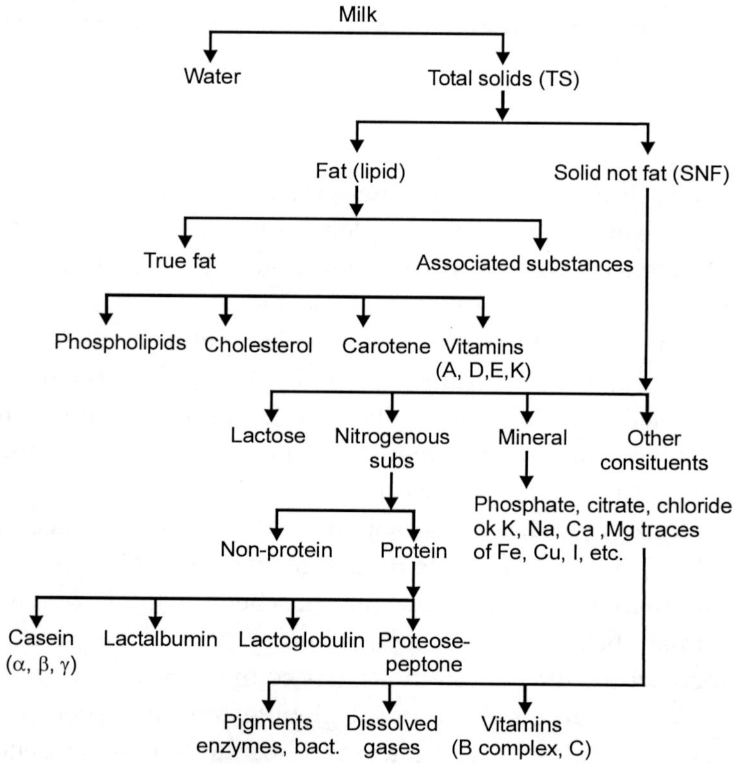

Other Components in Milk

Somatic cell: Milk always contains leukocytes cells, also known as somatic cells in cow milk. The concentration of leucocytes in milk varies with the species (human milk has relatively high somatic cell counts; cow milk from healthy glands has low somatic cell counts), infection status of the gland and stage of lactation.

Milk may also contain bio-active factors such as hormones and growth factors, enzymes, cellular proteins, and others.

Variation in the Composition of Milk

No two cows give milk of the same composition. The composition of milk of the same cow is not always the same. Of the different constituents of milk, fat content of milk, is most variable. After fat the other constituents vary in the followings order-protein, lactose, ash.

Table 2.1: The range of variation of cow's milk

Components	*Average per cent*	*Normal range of variation per cent*
Water	87.25	84.00 to 89.50
Fat	3.80	2.60 to 6.00
Protein	3.50	2.80 to 4.00
Lactose	4.80	4.50 to 5.20
Ash	0.65	0.60 to 0.80

Sourse: After Eekles

Gross Composition of Milk

Milk

Milk solids (12.68 %)   Whey protein (0.6%)

Table 2.2: Composition of milk (human, cow and buffalo)

Components	*Human (Per 100 ml)*	*Cow (Per 100 ml)*	*Buffalo (Per 100 ml)*
Proteins (g)	1.2	3.3	4.2
Fat (g)	3.8	4.1	7.0
Carbohydrates (g)	7.0	4.5	5.1
Ash (g)	0.21	0.72	0.82
Calcium (mg)	28.0	120	210
Magnesium (mg)	4.0	10.6	16.0
Sodium (mg)	17.0	50.6	-
Potassium (mg)	55.0	145	-
Phosphorus (mg)	14.0	82	128

Sourse: After Ling

Table 2.3: Chemical composition of milk of different species

Species	*%Water*	*%Fat*	*%Protein*	*%Lactose*	*%Ash*
Buffalo	84.2	6.6	3.9	5.2	0.8
Cow	86.6	4.6	3.4	4.9	0.7
Goat	86.5	4.5	3.5	4.7	0.8
Ewe	79.4	8.6	6.6	4.3	1.0
Sow	89.6	4.8	1.3	3.4	0.9
Human	87.7	3.6	1.8	6.8	0.1
Ass	90.0	1.3	1.7	6.5	0.5

Sourse: Outlines of Dairy Technology by Sukumar De (2003)

Water

- Water forms the largest fraction of milk and ranges from 80 to 90 per cent depending upon the species and breed.

- Water serves as a carrier for the other constituents of milk. A small amount of water in milk is hydrated to lactose and salts and also bound in the proteins and phospholipids.
- Its function is to hold the solids of the milk partly in solution and partly in suspension.

Milk Fat

- The fat exists in the form of small globules, which averages approximately 2 to 5 μ in size (diameter: cow 3 to 8 μ and buffalo 4 to 10 μ).
- This is an oil-in-water type emulsion.
- Fat globule membrane, a complex of phospholipids and proteins stabilises the fat emulsion, i.e. the membrane prevents the fat globules from coalescing and separating from one another and serves as some protection against the action of enzymes like lipase in causing the fat to become rancid. If held undisturbed, the fat globules tend to rise to the surface to form a cream layer, but the emulsion may be broken by agitation, freezing or heating.
- Milk fat is composed of a number of glyceride-esters of saturated or unsaturated fatty acids (make upto 98 to 99% of the milk fat) butyric, caproic, capryric and capric acids present in high proportions, are characterised by strong odours and flavour. The fatty acid content can be influenced by amount and feed consumed, stage of lactation and the breed of the animal.
- The size and arrangement of the globules affect the creaming ability of milk, the viscosity of cream, the ease of whipping cream and churning of cream to butter.
- The melting point of different fats differs widely; that of butter, a mixture of different fats, and ranges from 33 to 35.5°C. The normal body temperature of bovines is higher so the fat globules are present in liquid form in freshly drawn milk. Cooling of milk to a little below the melting

point of butter does not solidify the fat globules.

- Milk fat is quite bland in taste and imparts smoothness and palatability to fat containing dairy products.
- Milk fat is a rich source of energy and a good source of vitamin A and D.
- The colour of fat depends upon its carotene content and varies with the species, breed and feed of the animal. The yellow colour of cow milk is because of the carotene. Buffalo milk does not contain carotene.
- Milk fat also contains cholesterol (0.23-0.1 %) and phospholipids (lecithin, phosphatidyl serine, sphingomyelin, inositol and cerebrosides) some of which serve as antioxidants in prolonging the shelf-life of ghee.

Milk Proteins

- Milk proteins contain all essential amino acids in adequate amount and in balanced proportion. They are particularly rich in lysine and valine.
- Cow milk proteins possess high digestibility, biological value and growth promoting value. They are almost equal to human milk protein in infant nutrition.
- Three principal classes of proteins of milk are casein, α lactalbumin, and β lactoglobulin. α lactalbumin, and β lactoglobulin are known as whey protein or serum protein.

Casein

- It is a phosphoprotein, which is found in no other product than milk.
- The casein accounts for 80 per cent of total protein content.
- It exists in the form of calcium caseinate-phosphate.
- It is present in a colloidal suspension.
- Casein itself composed of α, β, γ, and κ fractions.
- Buffalo milk contains 44.5 per cent α casein, 52.4 per cent β casein, 3.1 per cent γ casein and 8 per cent κ casein as

against 54.5, 2.1, 6.4 and 10.5 per cent in cow milk casein. The casein content of buffalo milk is 4.3 per cent and that of cow milk is 3.0 per cent.

- Casein is coagulated by heat and may be precipitated by acid, rennet, alcohol, heat and concentration.
- The viscosity and white colour of milk are largely because of casein.

α Lactalbumin

- It constitutes about 3.5 per cent of the total nitrogen content of cow milk, in which it is held in solution.
- This is not coagulated by rennet or acid but is precipitated by heat.
- The average α lactalbumin content of milk of Indian cows is 0.4 per cent and of buffaloes is 0.383 per cent. The α lactalbumin content of colostrum is usually twice than in normal milk.

β Lactoglobulin

- It is dispersed in water of milk and coagulated by heat.
- Normal cow milk contains 10 per cent β lactoglobulin, although colostrum contains as much as 12.0 per cent.

Besides the above proteins milk also contains:

- Minor proteos peptone fraction
- Bovine serum albumin
- Immunoglobulins (IgG1, IgG2, and IgM), providing nonspecific humoral protective effect to the offspring against enteropathogenic microorganisms
- Lactoferrin, which deprives the enteropathogenic microorganisms to express their activity through its iron binding ability
- Lysozyme, which causes cell lysis and thereby provide bio-protective effect.

Milk Sugar or Lactose

- It occurs exclusively in mammalian milk.
- It is present in true solution and occurs in two forms, α and β.
- Chemically lactose is composed of one molecule of glucose and galactose.
- It is fermented by bacteria to yield lactic acid and other organic acids and is important both in the production of cultured milk products and in the spoilage of milk and milk products by souring.
- The average lactose content of cow and buffalo milk is 4.5 and 4.8 per cent respectively.
- Lactose contributes to the nutritive value of milk and milk products and is responsible for the texture and miscibility of some milk products.
- It also imparts colour and flavour to dairy products heated to high temperatures.
- Since lactose is not very soluble, it sometimes separates or crystallises in dairy products such as sweetened condensed products and ice-cream producing 'sandy texture' condition.

Enzymes

Enzymes are organic catalysts elaborated by living cells that catalyse many of the reactions involved in their functions. All known enzymes are proteins; some of them contain non-protein moieties called prosthetic groups which are essential to their catalytic activity, while others appear to consist of entirely of proteins.

Most enzymes exhibit a high degree of specificity acting only on surface substrates or lenkages.

Normal milk contains a number of enzymes. The milk enzymes constitute a very small portion of the total proteins in milk; never the less their characteristic catalytic activities make it possible to establish very precise analytical techniques for estimating them.

The principal enzyme natural to milk are catalase, peroxidase xanthine oxidase, alkaline phosphatase and acid phosphatase, amylases, protease, lipase and aldolase.

The milk enzymes are of considerable importance in certain aspects of milk technology.

The lipases, for example, are notorious for the production of hydrolytic rancidity in milk and dairy products. On the otherhand, the heat lability of certain enzymes furnishes the basis for important tests, for the extent of heat to which a sample of milk may have been subjected. Thus, the destruction of alkaline phosphatase is widely used as a test for the adequacy of pasteurisation.

Application of Enzymes in Dairy and Food Processing

"Enzymes are forcast to be a one billion dollar business by 2005. In India also, enzymes would be used extensively for diverse actions, resulting in novel enzymatic functions."

Over the past three decades, the manufacture of foods has greatly benefited through practical application of enzyme technology to food processing. Some common examples where enzymes have been employed in food processing are:

Dairy Industry: Cheese-Technology; Lactose Hydrolysis in Milk and Milk Products.

Food Industry: Starch Hydrolysis; Baking Processes; Clarification of wine and Fruit Juices; Meat Tenderisation, etc.

Table 2.4: Sources of selected industrially important enzymes

Enzymes	*Sources*	*Industrial use*
Animal enzymes		
Rennet	Abomasum	Cheese Technology
Trypsin	Pancreas	Dairy/Health tonics
Chymotrypsin	Pancreas	Dairy/Health tonics
Lipase	Pancreas	Lipid/Cheese Technology
Catalase	Liver	Egg/Processing

Contd...

Contd...

Plant enzymes		
Papain	Paio paio Latex	Meat Processing
Ficin	Fig Latex	Profecic Industry
Bromelain	Pineapple Latex	Brewing
Lipoxynase		
/3-gluconase	Soyabeans	Baking
α-Amylase		
/3-Amylase	Malted Barley	Brewing
Yeast enzymes		
Lipase	Candida	Butterfat Processing
Invertage	Saccharomyces	Confectionary
Lactase	Klyveromyces	Dairy Technology
Raffinase	Saccharomyces	Sugar Refining

Vitamins

- Milk is rich source of vitamin, such as thiamin and riboflavin, vitamin A (carotene), D, E and K.
- It also contains other B vitamins like niacin, pantothenic acid, biotin, folic acid and Vit.B_{12}, but it is a poor source of Vit. C.
- The contents of these vitamins in milk are affected by type of feed consumed by cow, stage of lactation, management and also by genetic factor.
- Sometimes the amount of vit. A in milk is artificially increased by oral administration of tocopherol.

Mineral

- Milk is an important source of calcium, potassium, phosphorus, and magnesium.
- It also contains very small quantities of iron, copper, iodine, cobalt, nickel, fluorine, manganese, and silica.
- The potassium and chloride are in solution whereas; phosphorus, calcium, and magnesium are partly in solution and partly in colloidal suspension.

- These minerals mainly remain as inorganic salts in milk, although some of them are in organic combination.
- Minerals constitute less than of 1 per cent of milk and they influence the heat stability and clumping of fat globules after homogenisation.

Nutritive Value of Milk

Milk has long been recognised as the most complete single food available in nature for the maintenance of health and promotion of growth of the mammals. The completeness of milk as food is appreciated when it is seen that it serves in the newborn animal as the sole substitute at least for a period, for the prenatal nourishment that the foetus was receiving through the placental circulation.

Milk is an almost ideal food and considered to be a perfect food. It has high nutritive value. It supplies body-building proteins, bone forming minerals, and health giving vitamins, and furnishes energy giving lactose and milk fat.

- *Water:* About 84 to 88 per cent of milk is water that holds various solids in suspension and helps to maintain body temperature.
- *Protein:* Milk protein, i.e. lactalbumin and lactoglobulin are a very good source of all essential amino acids. These are superior to vegetable protein. These proteins are required for the repair and maintenance of body tissues.
- *Fat:* Milk fat is superior to other animal fat because it contains essential fatty acids like linoleic acid, arachidonic acid, etc. and serves as a rich source of energy. It also enhances the flavour of milk, thereby enhances the consumer's acceptability of food.
- *Carbohydrate:* The main carbohydrate in milk is lactose, which makes it superior over other carbohydrates. It facilitates assimilation of calcium in the body and helps to check the growth of proteolytic bacteria by creating an acidic medium in the intestine.

Vitamins

Vitamins are required in minute quantities for metabolism of food and enzyme action. Vit. A is responsible for growth, resistance to infection, proper eye function, pregnancy, and lactation. Vitamin B is essential for growth, reproduction, healthy skin, and proper nervous functions. Vitamin C is called ascorbic acid and is required for teeth, gums, capillary action, and respiratory system. It prevents infection and skin diseases. Vitamin D is required for proper utilisation of calcium, and phosphorous. It is also an anti-rickets factor. Vitamin E is called anti-sterility factor. Vitamin K is blood-clotting factor. A proper balance of vitamins is required in the body and the system is called Polyvitaminosis. Want of vitamins, on the other hand, causes disease and restrict growth and is called avitaminosis.

Trace Elements

In addition to the elements that occur in milk in relatively large proportions (K, Ca, Na, Mg, P, Cl and S) there are large numbers of elements usually measured in ppm or μg per liter and referred to as trace elements. The traceelements in milk generally come from the feeds eaten by the animals. Colostrum invariably contains a higher proportion of trace 'elements than normal milk. Copper is essential in the formation of haemoglobin; a deficiency produces anaemia. Iodine has physiological function as a constituent of thyroxin and a deficiency cause an enlargement of the thyroid, the enlarged gland being known as goiter.

Milk is also a 'mini-bank' for certain trace elements, which include a long list of metals, metalloids like, arsenic, boron, and silicon and the halogens. The metabolic pathway find in their entry from fodder to udder and also from the radionucleids in milk. The metallic containers used during milk storage often become the donor for trace elements in milk.

Copper and iron has got catalytic function in the oxidation of milk lipid. Iron also co-exists with some milk enzymes associated with the milk-fat globule membrane.

Biological Value of Milk

Both cow, buffalo milks have a high biological value, digestibility, and protein efficiency.

	Biological value	*True digestibility*	*Protein efficiency*
Cow	86.90	91.96	2.60
Buffalo	84.04	88.86	2.38

Energy Value of Milk

The energy value of milk will vary with its composition. On average, cow milk furnishes 75 cal/100 g and buffalo milk 100 cal/100 g.

The energy giving constituents and their individual contributions are as follows:

- Milk fat — 9.3 cal/g
- Milk protein — 4.1 cal/g
- Milk sugar — 4.1 cal/g.

CHAPTER 3

Factors Affecting Composition of Milk

Milk from different sources show a variation. They vary individually as well as collectively. The variation is always greater in small herds than in large ones. Generally, the greatest daily variation found in case of milk is in fat, and then comes protein followed by ash and sugar.

The factors affecting composition of milk are discussed below:

i. *Species:* Composition of milk varies from species to species, which have been presented in the Table 2.3 showing composition of milk from different animals.

ii. *Breed:* Breed also affects the composition. Generally, high milk yielding breed produce lower fat per centage. Cattle are classified as milch, draught, and dual-purpose breed on the basis of milk production.

iii. *Individuality:* Each animal tends to yield milk of a composition that is characteristic of the individual. Animals of a particular breed, reared under uniform environmental condition, produce milk with different composition due to genetic variation amongst them. This variation in milk is more pronounced in case of milk fat and to some extent in SNF.

iv. *Age of the animal:* The age of an animal affects the milk yield and also the composition of milk. With the advancement of the age, the milk yield decreases and the milk composition changes. The fat per centage of milk increases slightly with the advancement of age upto the third lactation and after that it remains practically unchanged until age begin to affect the animal.

v. *Health of the animal at parturition:* Health of the animal at the parturition plays an important role in controlling the milk composition. The animal with good health at that period produces more milk with higher fat content. On the contrary, the weak or sick animal produce milk with less fat and some times with abnormal odour and poor heat-stability. The animals highly fatty at parturition tend to produce milk with higher fat per centage.

vi. *Stage of lactation:* The first secretion (colostrum), which lasts for 3 to 5 days after calving is quiet different from normal milk, having a higher specific gravity due to its higher total solid content than normal milk. It is reddish yellow, slimy, viscous, bitter in taste, and possess abnormal odour. As the lactation progress, the colostrum changes into normal milk. Upto 15 days of parturition, the SNF content decreases and during the last three months of lactation the fat and SNF content increases continuously. Size of fat globule is fairly constant during 5th to 7th month and smaller during last month of lactation period. The level of lactose in milk remains quiet constant throughout the lactation period. The protein content increases near the end of the lactation period.

vii. *Season:* Both the fat and SNF content of milk vary slightly but significantly throughout the year. This variation is due to availability of green fodder and stage of lactation. In monsoon and winter fat percentage is lower than summer because of consumption of green fodder during first two season.

viii. *Stages of milking:* As the milking progresses, the fat content in milk increases. The stripping, i.e. the milk drawn at the end of the milking contains higher fat than the foremilk. Therefore, the animal should be milked completely to avoid variation in milk composition.

ix. *Interval of milking:* The longer milking interval results in higher milk yield with lower fat content. Morning milk differs significantly in milk composition than evening milk and generally, evening milk contains higher fat than morning milk.

x. *Irregularity of milking:* Frequent changes in the time of milking and milking interval affect the milk composition.

xi. *Feed:* The animals maintained with concentrate feed, produce milk with higher fat and SNF. Oil cakes and cottonseed meal directly affect the fat content. Green fodder may cause yellowish colour in milk. The lactose content and mineral are not affected by the feed.

xii. *Oestrus cycle:* Animals produce abnormal milk during heat period. Generally they produce milk with less fat content when they are in heat.

xiii. *Exercise:* Exercise before milking results in less milk but higher milk fat than normal milk.

xiv. *Disease:* Diseases, particularly mastitis or any injury to udder affect the milk composition adversely. In case of mastitis, milk contains higher chlorides and lower lactose with a bitter taste and offensive odour. Such milk has higher albumin and lower casein content along with poor heat-stability.

xv. *Milkers:* Any change in milker adversely affects the quantity and quality of milk due to incomplete letdown of milk by the animal.

xvi. *Excitement:* A transient fluctuation in milk yield and composition is likely to occur during period of excitement.

xvii. *Administration of drugs and hormones:* Generally, milk yield, fat and SNF contents in milk are adversely affected by administration of certain drugs. Injection or feeding of hormones increases both the yield and fat per centage. Antibiotic may affect the desirable changes in milk by preventing starter organism.

CHAPTER 4

Physical and Chemical Properties of Milk

Physical and Chemical Properties of Milk

The milk constituents exist partially as a solution and partially in colloidal state. In general milk is a emulsion of fat in water with some of the protein and salts as binding material or interface substances in colloidal suspension, while milk sugar, some proteins, and salts are soluble in water.

PHYSICAL PROPERTIES

Colour

Milk is opaque fluid varying from white to yellowish white. Colour is due to: a. Reflection of light from casein particles render whiteness. b. Refraction of light through fat particles renders whiteness. c. Yellow pigment carotenes render milk yellow.

White milk: Buffalo, goat, or sheep milk.

Yellow milk: grass-fed cows milk due to carotene.

White yellowish milk: Stall-fed cows milk.

Bluish white milk: Skim milk or milk from which fat has been removed. The bluish color appears due to another milk constituents, i.e. riboflavin.

Greenish whey: The whey which is available by removing milk protein by curdling still contains some fat and riboflavin. The effect of refraction and colour combination makes it green.

Flavour

Normal flavour of milk is the combination of taste and smell. The chlorides and lactose are mainly responsible for taste. The ratio of chloride and lactose, i.e. per cent of chlorides/ per cent of lactose × 100 or per centage ratio is called chloride-lactose number. When the value is, i.e. number is 1.2 to 2.0 the milk is good; a higher value or number means salty. A faint animal flavour is found in normal milk but if pronounced other flavours are present the same are called off flavour.

Taste

Milk is sweet in taste. The sweetness is attributed to lactose. The chlorides, which are salty or bitter, also control a part of milk taste. More lactose makes milk sweeter.

Specific Gravity

The specific gravity of water is 1.00. Milk contains constituents whose specific gravity is greater than that of water and hence greater than 1. The average specific gravity of milk is generally determined by lactometer. The specific gravity is used to judge the purity of milk.

The specific gravity of milk means the weight of given volume of milk compared with the weight of the same volume of water at the same temperature. The specific gravity of milk is determined at 15.6°C by using a lactometer. The common types of lactometer are Zeal, Quevenne, etc. Using the lactometer reading, the total solids are calculated by the following formula: per centage of total solids =CLR/4 + 1.2F + 0.14, Where, F = per centage of fat; CLR = corrected lactometer reading (OLR + 0.1 (T –84); OLR = original lactometer reading, and T = temperature of milk. Specific gravity of milk = 1+ CLR/1000. Pycnometer or a hydrostatic balance can also determine specific gravity of milk.

The proportion of its constituents influences the specific gravity of milk. Fat being the most variable ingredient of milk has maximum influence on the specific gravity of milk. Increase in the fat content of milk results in lowering of specific gravity and decrease in the fat content of milk causing increase in specific gravity. Generally, the specific gravity of cow milk ranges from 1.028 to 1.032 and that of buffalo milk from 1.030 to 1.034. The specific gravity of skim milk should always be higher than that of whole milk from which it is obtained. The specific gravity varies with temperature being lower at higher temperature and vice-versa. Addition of water to milk causes decrease in the specific gravity, while addition of solids raises the specific gravity of milk. Determination of specific gravity of milk is a quick test to detect the adulteration of milk with water and/or other solids.

Density

Density of a substance is its mass (weight) per unit volume. The density of milk is taken with the hydrometer at 20°C. The method of taking the hydrometer reading is the same as in lactometer reading. The density of milk is used in conjunction with the fat test to estimate the content of total solid. The total solids in milk are calculated by using the following formula: per centage of total solids = 0.25D + 1.21F + 0.66, where, D = 1000 (d–1) , d = density of sample of milk at 20°C and F = fat per centage of sample.

Refractive Index

The refractive index of milk is sometimes used to indicate adulteration, specially watering. The relation between solid contents of milk and refractive index is linear and the contribution of different constituents is additive. The refractive index of milk at 20°C ranges from 1.344 to 1.3485, that of water being 1.33.

Viscosity

The viscosity of a substance refers to its resistance to flow. The viscosity of milk depends upon its composition, temperature, and physical state of its colloidally dispersed substances, including milk fat. The viscosity decreases with increase in temperature, i.e. it has a negative temperature co-efficient. At normal temperature, viscosity of milk varies from 1.5 to 2 centipoises. The viscosity of milk is determined by viscosimeter. The viscosity of milk is always higher than the viscosity of water due to the presence of dissolved solids in the milk. Homogenisation increases the viscosity of milk.

Surface Tension

Milk has a surface tension of 50 dyne per cm at 20°C. Its surface activity is related to proteins, fat, phospholipids, and fresh fatty acids present in it, and also the temperature of milk. Surface tension decreases as the temperature increases, at 60°C it is about 40 to 45 dynes/cm. The presence of fat lower surface tension than that of skim milk and that of cream is still lower. Development of acidity and churning lower the surface tension of milk.

Freezing Point

The temperature at which, at a given pressure, a material is in equilibrium as both a solid and as liquid is called the freezing point and at this temperature the liquid phase may freeze or crystallise and the solid phase may melt or liquefy. Milk freezes at temperature slightly lower than that of water due to the soluble constituents (lactose, soluble salts), which lower or depress the freezing point. Freezing point of milk of cow or buffalo ranges from –0.535 to –0.55°C with an average of –0.545°C, which can be raised to 0.006°C by adding 1 per cent water to milk. Freezing point of milk is considered very useful in detecting adulteration of milk with water because it is one of the most constant physical properties of milk despite the variations in the normal. The freezing point of milk is determined by cryoscopy. The lactose, minerals and dissolved carbon dioxide are the principal contributors to the freezing point of milk.

Boiling Point

The boiling point of any liquid is the temperature at which at the given pressure, the material is in equilibrium both as a liquid and as a gas. The presence of dissolved substances increases the boiling point of a solution. The boiling point of milk varies from 100.17 to 101°C. This is slightly higher than that of water due to dissolved substances in milk.

Specific Heat

Specific heat of a substance is the number of calories required to raise 1°C temperature of 1 g of a substance. Usually specific heat of milk ranges from 0.92 to 0.93 with highest value of 0.938 at 15°C. It changes with temperature. The cost of cooling milk is related to its specific heat.

Co-efficient of Thermal Expasion

Milk expands when heated and contracts when cooled, and the ratio of increase in volume per degree of increase in temperature is known as "co-efficient of expansion". This property is of great importance in a dairy because, when filling in vats is done with cold water for subsequent pasteurisation, one has to take into account the expansion that will take place on heating of milk.

Electrical Conductivity

In milk, the fat and colloidally dispersed substances obstruct the ions in their migration and decrease conductivity. The electrical conductivity increases with increase in temperature. Values for normal milk range from 0.003 to 0.005 mhos/min values above 0.006-mhos indicate general secretion disorder. It can be used to detect added neutralisers and abnormal milk.

Thermal Conductivity

It is important because it determines how fast the milk is heated or cooled. At 30°C thermal conductivity of milk is 13.5×10^4 cal cm-1 S-1 C-1. It increases as temperature increases and it decreases as concentration level increases. At a given temperature, the higher the fat contents the lower the thermal conductivity.

Heat Stability of Milk

The heat stability is normally defined as the length of time required to induce coagulation at a given temperature or the temperature required to induce coagulation in a given time. It is very important property for the processing of milk. Casein exists as micelle and takes part to a notable extent in determining heat stability of milk. The salt balance of milk governs the heat stability of milk. The effect of added citrates, phosphates, and calcium ion on heat stability tends to play an import role. In the manufacture of concentrated milk preparation such as evaporated and condensed milk and milk powder, heat stability of milk is of utmost importance.

CHEMICAL PROPERTIES

Chemical Reaction

Fresh milk is amphoteric in nature due to the presence of different amino acids in proteins. Amino acids consist of NH_2^{++}

as well as $COOH^-$ groups, which show alkaline as well as acidic reactions.

Acidity of Milk

Milk shows certain acidity as determined by titration with an alkali in the presence of an indicator. This is called titrable acidity of milk. The acidic nature of freshly drawn milk is due to the presence of phosphate, citrate, casein, and dissolved carbon dioxide. This acidity of milk is called natural acidity. Natural acidity in milk is considered important from the heat stability point of view. When milk is kept under atmospheric conditions for sometime, an increase in acidity known as developed acidity, which is due to production of lactic acid as a result of microbial activity on lactose of milk is observed. Hence the titrable acidity is expressed as a per centage of lactic acid. The natural acidity of milk may range from 0.13 to 0.21 per cent. The acidity of fresh milk varies with species, breed, individuality, stage of lactation, health of the animal, etc.

pH of Milk

The pH of normal, fresh milk ranges from 6.5 to 6.7, i.e. the milk is slightly acidic. Higher pH (alkaline) of milk indicates that the animal is suffering from udder infection, like mastitis and this milk have abnormal odour, taste, and poor heat-stability. Lower pH value (below 6.5) of milk is considered to be due to added colostrums or bacterial deterioration.

Buffering Action of Milk

Fresh milk acts as a complex buffer because it contains several substances, which contribute to this property of milk. Certain constituents namely, carbon dioxide, proteins, citrates, and a number of minor constituents are responsible for buffering capacity of milk. This property of milk is important from the curdling and from heat stability point of view.

Oxidation-reduction Potential

The oxidation-reduction potential or Redox potential of milk is designated by the symbol Eh of milk which normally falls within the range of +0.2 to 0.3 volt. Certain factors such as heat treatment, bacterial activity, and contamination with trace metal, particularly with copper affect the Eh of milk. Methylene blue reduction test of milk is based on the principle of lowering of the potential through use of available oxygen by the microorganisms.

CHAPTER 5

Clean Milk Production

Milk is a potential carrier of pathogenic and non-pathogenic microorganisms which can cause spoilage and health hazards if the milk is consumed before they are not destroyed. Since most of the organisms gain entry into the milk after it leaves the udder, their population could be fairly controlled by proper care during production and handling. In India, most of the milk is produced in rural areas where sanitary conditions are far from satisfactory. All possible efforts should be made to secure milk with low bacterial count, free from visible dirt, and pathogenic microorganisms. Hence, proper care must be taken regarding preservation of milk, protection of milk constituents, protection against high temperature, and natural calamity.

Following factors should be borne in mind while considering clean milk production:

1. *Health of the animals:* Milk should be drawn only from healthy animals. Animals should be free from infectious diseases. Udder should neither be affected by mastitis nor should there be any wound. Those animals, which are either sick or have infection of any kind, should be segregated and either should not be milked or milk produced by them should be handled separately.

Table 5.1: Sources of contamination of milk and their control

	Source of contamination	*Control measure*
1.	Interior of the udder	a. Check for mastitis b. Discard premilk
2.	Exterior of the cow, particularly udder and flanks	a. Wash and wipe udder b. Clip the udder and flanks c. Dry milking d. Use small-top milk-pail
3.	Barn air and dust	a. Keep milk covered
4.	Flies and other vermin traps	a. Eliminate breeding places b. Fly control with fly sprays, fly repellent, etc.
5.	The milker	a. Clean habits b. Dry milking
6.	Utensils before use.	a. Clean, sanitise and dry

Source: Outlines of Dairy Technology by Sukumar De (2003)

2. *Cleaning of animals:* The animals should be given a bath before milking. If it is not possible, the udder and the adjoining parts should be washed thoroughly. Teats should be cleaned with disinfectant solution and dried with a clean cloth dipped in such solution of chlorine, bleaching powder, or potassium permanganate. Clipping of hairs in udder region, grooming and brushing the animal before milking will help to minimise the bacterial population in milk.

3. *Using small-topped milking pails*: Normally 50% of the dirt, which falls in the milk, could be prevented by using small-topped containers. Use of hooded milking pails are helpful in preventing the dirt, coming to milk during milking as compared with open-top milking pail. Milking pails should be washed immediately after milking is over. They should be washed with clean water followed by detergent washing and finally with clean water. Use of soda ash or

surf will help to remove milk solids from the container. The milking pail, after proper washing should be sterilised either by heat treatment or by keeping it under the bright sunlight. Chlorine or bleaching powder solution of proper strength can also be used for rinsing the milk pails, which will effectively control bacterial population.

4. *Health and cleanliness of milker*: The milker should be free from contagious diseases and should be in sound health. It is important to understand that the health of the milker is as important as the health of the animal. Milker should clean the hands before milking with clean water and the hands should be wiped off with a cloth previously dipped in disinfectant solution. Smoking as well as chewing of tobacco should be avoided just before and during milking.

5. *Milking barns*: It is essential that the place where milking is done is kept clean. The barns should be well ventilated and there should be adequate lighting arrangements. Cleaning of barns should be done at least one hour before the milking starts. Accumulation of dung and urine in the barn should be avoided. It should be protected from flies, mosquitoes, rats, cockroaches, and birds, etc. If possible white washing of the walls, etc. should be done periodically.

6. *Cleaning of utensils*: From the milking pails milk is generally transferred either in milk cans or other containers. In order to minimise bacterial number in milk, all the containers with which milk comes in contact must be properly cleaned and sterilised. There should be no accumulation of milk solids in these containers and they should be properly covered.

7. *Straining of milk*: The process of straining helps to remove large size objects, which come to the milk from the

external sources. For straining the milk, either metallic strainer or muslin cloth be used. Use of cotton pads for straining milk become more effective. One should clearly keep in mind that proper cleaning and sterilisation of strainer as well as cloth before they are used, is a primary requirement for achieving the objectives of clean milk production.

8. *Cooling of milk*: Milk should be cooled to less than 10°C as quickly as possible after milking. This process will help to check the growth of microorganisms, which have gained entry into milk. Lower temperature check bacterial growth, therefore possible efforts should be made to reduce the temperature of milk after it is drawn from the udder of animal. Freshly drawn milk has a temperature of approximately 38°C, which is highly favourable for the bacterial growth.

CHAPTER 6

Bacteriology and Deterioration of Milk

Milk in the udder is normally microorganisms free; however, microbes may by present in mammary gland of unhealthy animals. As it leaves the udder, many microbes may enter the milk. These microbes bring the changes in flavour and appearance of the milk. Most important microorganisms in dairy are bacteria, yeast, mould, and virus where the first one predominates. Milk is a very good medium for microbial growth, so as soon as they enter the milk, they grow, multiply, and produce several biochemical changes, which are considered undesirable in milk.

In milk and milk products, the spherical and cylindrical forms of bacteria are predominant. Spore forming bacteria cause great problem in dairy because of their resistance to pasteurisation and sanitisation procedure.

Source of Bacterial Contamination

There are several sources of bacterial contamination in milk. Out of these, most important sources are briefly discussed here.

1. *Udder:* As udder harbours bacteria, first few strips of milk contain a large number of bacteria. Milk of healthy cows mostly contains non-pathogenic bacteria whereas milk

from diseased cow suffering from TB, brucellosis, FMD, anthrax, mastitis, etc. is infected with germs of these diseases.

2. *Milkers:* Milkers must be clean and healthy as because milkers may directly transmit some milk-borne diseases to the milk.
3. *Cow's body:* In case of unclean cows, the manure, soil, hair, straw from tail, udder, and flank may fall into milk during milking.
4. *Stable air:* Milking operation should be done in a dust free atmosphere because stable air is an important source of contamination.
5. *Utensil:* Dairy utensils like pails, cans, bulk containers, etc. should be made of stainless steel or aluminium alloy, free from pores, rust, sharp angles, seals, or crevices to enable easy and thorough cleaning, otherwise they will be a potent source of contamination.

Mode of Transmission of Microbes in Milk

The microorganisms may enter the milk in the following ways:

1. *Infection of milk directly from the cow:* The microorganisms enter the milk through udder or faecal contamination; e.g. *Mycobacterium tuberculosis, Brucella melitensis, Bacillus anthracis.*
2. *Infection from men to animal to milk:* The microorganisms essentially cause human disease but can become established in the udder, e.g. *Hemolytic streptococci, Str scarlatina, Corynebacterium diphtheriae.*
3. *Direct contamination of milk by human being,* e.g. *Salmonella typhosa, H. streptococci, Shigella dysenteriae, Entamoeba histolytica,* etc.
4. *Indirect contamination of milk by human being:* Pathogenic organisms enter the milk through contaminated utensils,

water supply, insects, etc. e.g. *Salmonella typhosa, Shigella dysenteriae*, etc.

CLASSIFICATION OF BACTERIA PRESENT IN MILK

Based on Pathogenicity

Bacteria are classified into two categories.

1. *Pathogenic:* Bacteria, which produce disease in the body, are pathogenic bacteria. Therefore, they should be destroyed before consumption of milk.
2. *Non-pathogenic:* These bacteria do not cause disease but are capable of causing spoilage of milk. Some of these are also beneficial in dairy industry for product manufacturing.

On the Basis of Temperature for Growth

1. *Psychrophilic:* They grow within the temperature range of 3 to 20°C (optimum is 7°C), e.g. *Pseudomonas fragi, Pseudomonas fluorescens.*
2. *Mesophilic:* They can grow at temperature range of 20 to 50°C (optimum is 37°C), e.g. *Streptococcus lactis, Strep. cremoris.*
3. *Thermophilic:* They grow at temperature above 50°C, e.g. *Bacillus calidolactics, Lactobacillus thermophilus*

On the Basis of Bacterial Action on Milk

1. *Lactose fermenting bacteria:* They convert the lactose of milk into lactic acid. They are of two types:
   i. Homofermentative bacteria are those which produce only lactic acid and small quantity of carbon dioxide, e.g. *Streptococcus lactis, Strep. cremoris.*
   ii. *Heterofermentative bacteria* are those, which produce certain other substances apart from lactic acid, e.g. *Lactobacillus sp. Leuconostoc sp.*

2. *Proteolytic bacteria:* These bacteria hydrolyse protein which results in increase in pH and coagulation of casein. The enzyme rennin from proteolytic bacteria causes coagulation without acid production and this is known as sweet curdling. This is followed by proteolysis, which causes alkaline reaction and bitter taste, e.g. *Bacillus subtilis, B cereus, Pseudomonas putrificiens,* etc.
3. *Lipolytic bacteria:* These bacteria produce enzyme lipase which hydrolyses milkfat into glycerol and fatty acids. Some of these fatty acids impart a rancid flavour in milk, e.g. *Pseudomonas fragi, Ps fluoroscence,* etc.

Factors Affecting Growth of Microorganisms

The microbial growth is affected by following factors:

i. *Nutrients:* Milk and its product are rich in protein, fat, carbohydrate, and minerals, which furnish all the food requirements of microorganisms.

ii. *Moisture:* Except some condensed and dehydrated milk product, milk contains adequate water, which is an absolute requirement for microbial growth.

iii. *Oxygen:* Aerobic bacteria grow only in the presence of oxygen, while anaerobic bacteria do not require oxygen for their growth. Milk contains both of these bacteria and facilitates their growth.

iv. *Acidity or pH:* Normal pH of milk is 6.4 to 6.6 and most common bacteria prefer a pH from 5.6 to 7.5.

v. *Preservative:* Preservatives like formalin, boric acid, benzoic acid, hydrogen peroxide, etc. destroy or check the growth depending on their concentration.

vi. *Light:* Direct sunlight and UV light are used to destroy the microbes. Bright light checks the microbial growth.

vii. *Temperature:* Each microorganism has its own optimum, maximum and minimum temperature of growth. It is an important means to control the microbial growth.

viii. *Concentration of salt and sugars:* High concentration of salt or sugar in product checks bacterial growth.

Sequence of Fermentation

When microorganisms are allowed to grow, the changes in milk take place in the following sequence:

1. *Acid production:* The first important change due to bacterial activity is the transformation of lactose into lactic acid. Some lactobacilli continue to produce acid until it reaches to approximately 4.0 per cent.
2. *Neutralisation:* The growth of acid producing organisms is checked when sufficient acid has already been produced. Their metabolic products reduce acidity and neutralise the medium of growth.
3. *Putrefaction:* Some proteolytic bacteria become active after neutralisation, decompose the substance completely and separate a clear liquid by acting on casein. The milk becomes unfit for human consumption.

Effects/Results of Microbial Growth in Milk

1. *Souring:* The lactic acid fermentation is most likely to take place in raw milk held at room temperature. Souring also take place at different temperature.

Temperature	*Organism*
10°C–37 °C	*Streptococcus lactis*, growth of some coliforms enterococci, lactobacilli and micrococci.
37°C–50°C	*S. thermophilus, S. faecalis, L. bulgaricus*
Above 50°C	*L. thermophilus.*

2. *Gas production:* Gas production by bacteria is accompanied by acid formation and is undesirable in milk and milk products. The chief gas formers are coliform bacteria, *Clostridium* spp, gas forming *Bacillus* spp. Those yield both hydrogen and carbon dioxide. The production of gas cause so called "stormy fermentation" of milk.

3. *Proteolysis:* The hydrolysis of milk proteins by microorganisms accompanied by the production of a bitter flavour due to release of some protein splitting product—peptides. Species of Micrococcus; *Alcaligenes, Pseudomonas, Proteus, Flavobacterium*, and *Serratia.*

   The types of change produced by proteolytic microorganisms:

   i. Acid proteolysis, in which acid production and proteolysis occur.
   ii. Proteolysis with little acidity or even alkalinity.
   iii. Slow proteolysis by intracellular enzymes of bacteria after their autolysis.
   iv. Residual proteolytic activity of heat stable proteinase.

4. *Ropiness:* Bacterial ropiness is caused by slimy capsular material from the cells, usually gums or mucins, producing rope like thread in milk and ordinarily develops best at low storage temperatures.

Types of ropiness	*Organism*
Surface ropiness	*Alcaligenes viscolactis, Micrococcus freudenreichii*
Ropiness throughout the milk	*Enterobactor aerogenes, Klebsiella oxytoca, streptococcus lactis var hollandicus.*

5. *Changes in milk fat:* Some bacteria, yeast, and moulds may decompose milk fat causing oxidation or hydrolysis and produce pronounced odour-rancidity. The following changes in the milk fat take place:
   i. Oxidation of the unsaturated fatty acids
   ii. Hydrolysis of the butter fat
   iii. Combined oxidation and hydrolysis producing rancidity.
   Pseudomonas spp; *Alcaligenes* spp, *Bacillus* spp, and *Micrococci* spp. are found responsible for causing such changes.
6. *Alkali production:* The group of alkali producers includes bacteria, which cause an alkaline reaction in milk without any evidence of proteolysis. The alkaline reaction may result from the formation of ammonia, as from urea, or of carbohydrates, as from organic acids such as citric acid. Alkali formers are *Pseudomonas fluorescens* and *A viscolactis*.
7. *Colour changes:* The colour of milk is affected by its physical and chemical composition. The colour may be due to the surface growth of pigmented bacteria, or moulds in the form of a scum or ring or may be present throughout the milk. Pseudomonas spp are mostly involved in the colour change situations.
8. *Flavour changes:* Normal freshly drawn milk should have clean sweet flavour, free from any abnormal flavours. Some abnormal flavours, e.g. sour flavour, bitter flavour, fruity flavour, caramel flavour may be produced by some microorganisms *S. lactis, Coliforms, Alcaligenes* spp, *Pseudomonas* spp, cause such changes.

Table 6.1: Action of microbes on milk at a glance

Microbes	*Action*	*Result*
Streptococcus lactis	Souring	Lactose to lactic acid, casein precipitation
Lactobacillus casei	Cheese ripening	Controls intestinal fermentation
E. coli	Souring and gassiness	Lactic acid and gases affects cheese ripening
Bacillus subtilis	Proteolysis	Off/putrid flavour
Alcaligenes viscosus	Ropiness	Ropy milk
Streptococcus liquefacieus	Bitter flavour	Bitter flavour to cream and butter
Bacillus subtilis	Sweet acid	Curd formation
Streptococcus paracitrovorus	Attacks citric acid	Add flavour to curd

Table 6.2: Flavour changes in milk caused by microorganisms

Flavour	*Causative organisms*
Sour or acid flavour	*Streptococcus lactis, Coliforms, Leuconostoc* spp *Clostridium* spp
Bitter flavour	*Alcaligenes viscolactis, Micrococcus freudenreichii, Enterobacter aerogenes; Klebsiella oxytoca; streptococcus lactis var hollandicus,* some strains of *asporogenous* yeasts
Burnt or caramel flavour	*Streptococcus lactis var maltigenes*
Barny flavour	*Enterobacter oxytocum*
Turnip like flavour	*E. coli, Pseudomonas fluorescens*
Malt flavour	*Yellow micrococci*
Fruity flavour	*Pseudomonas fragi*
Potato like flavour	*Pseudomonas mucidolens*

Table 6.3: Colour changes in milk by microorganisms

Colour	*Causative organisms*
Blue milk	*Pseudomonas syncyanea* (bluish gray), *streptococcus lactis* (dark blue)
Yellow milk	*Pseudomonas synxantha, Flavobacterium*
Red milk	*Serratia marcescens, Brevibacterium erythrogenes, Micrococcus rosens*
Brown milk	*Pseudomonas putrefaciens, Pseudomonas fluorescens*

CHAPTER 7

Pasteurisation

PRINCIPLE OF PASTEURISATION

It is the process of heating every particle of milk or milk product, in properly designed and operated equipment to specified temperature and holding at that temperature for specified period of time followed by immediate cooling and storing at low temperatures.

METHODS OF PASTEURISATION

Batch Pasteurisation

Under batch process the milk is heated to 63°C for 30 min. in a double-jacketed vat. Heating and cooling is done by spraying or circulating hot water/steam or chilled water between the inner and outer jacket of the vessel. The process is called low temperature long time (LTLT) method.

Merits

i. Suitable for small quantity (200-1000lt)
ii. Low initial cost
iii. No milk loss
iv. Simple process

Demerits

i. Long time process
ii. Growth of thermophilic organism may be encouraged
iii. CIP is not convenient

High Temperature Short Time Pasteurisation

High temperature short time (HTST) treatment for pasteurisation of milk refers to heating every particle of milk in a continuous flow to a minimum of 72°C for atleast 15 seconds, followed by immediate cooling to 4°C. The complete process of preheating, heating, holding, pre-cooling, and chilling is completed in a plate type heat exchanger mounted on a compact frame with inter connected sections to make the process continuous.

Merits

i. Ideal for large scale handling of 5000 l/hour or higher
ii. Automatic process, less labour requirement
iii. Less chance of contamination
iv. Milk is readily available for further processing

Demerits

i. High initial cost
ii. High refrigeration cost
iii. Chances of milk loss in pipe lines
iv. Requires an experience operator

Vacuum Pasteurisation

This process is also called vacreation where mixing of milk with pressure steam immediately followed by sudden and spontaneous expansion into a finely divided state with a chamber maintained at reduced pressure. This is more suitable for cream pasteurisation for butter making.

Flash Pasteurisation

The temperature of milk is raised to 80°C for practically no holding period to destroy all the pathogenic bacteria with less reduction in the nutritive value of milk.

Ultra Pasteurisation (Uperisation)

This is itself a shortened form of the word ultrapasteurisation that has been developed in Switzerland. In this process, milk is heated with direct steam up to 150°C for a fraction of a second. This is a continuous process.

Flow Diagram

Reception of high quality raw milk—clarified immediately—chilled and stored—fore warmed to 50°C—deaerated (vacuum treatment)——pre-heated to 80 to 90°C——heated in uperisation chamber (high pressure, 150°C for Ä to ¾ second—evaporation of moisture in expansion chamber——cooler——storage.

Ultra High Temperature Pasteurisation

Milk is heated to 135-150°C for practically no holding time either by direct steam injection, swept surface heat exchanger, or turbulent heat exchanger. This process maintains colour, flavour and nutritive value of milk.

Stassanisation

The milk is heated to 74°C for 7 second by passing it through two water heated pipes through a narrow space of 0.6 to 0.8 mm and immediately cooled to 4 to 5°C.

Table 7.1: Time-temperature relationship in pasteurisation

Process	*Temperature (°C)*	*Time (s)*
Low temperature long time pasteurisation	63	1800
High temperature short time pasteurisation	72	15-20
Ultra pasteurisation	115-130	2-4
Ultra high temperature pasteurisation	135-150	1-6
Vacuum pasteurisation (Vacreation)	1st stage: 90-90 2nd stage: 71-82 3rd stage: 43	10
Stassanisation	74	7

Effect of Pasteurisation on Nutritive Value

Pasteurisation may affect the nutritive value of milk adversely. Heating process destroys certain vitamins like B_1 (10 %), C (5-10 %) in holding method. Some important minerals such as magnesium and calcium may be precipitated. This process reduces the cream line which is an important criteria for better consumer-acceptability.

Advantages of Pasteurisation

1. It makes the milk safe for human consumption by destroying the entire pathological microorganisms.
2. It increases the keeping quality of milk by destroying almost all spoilage microorganisms.

Disadvantages of Pasteurisation

1. Nutritive value of milk is to some extent adversely affected.
2. It affects the cream line.
3. It affects clotting of milk with rennet.
4. Increased milk cost due to processing charges.
5. Bacterial toxin in milk is not destroyed.
6. Bacterial spores cannot be destroyed.

HOMOGENISATION

Principle of Homogenisation

Homogenisation refers to the process of forcing the milk through a homogeniser with the object of sub-dividing the fat globules. The process is carried out to prevent creaming in stored milk by reducing the fat globule size to 1 micron or less. The process involves pumping of milk at 60 to 65°C and product at 75 to 80°C through a small passage at high velocity at a pressure ranging from 10 to 25 Psi depending on the product.

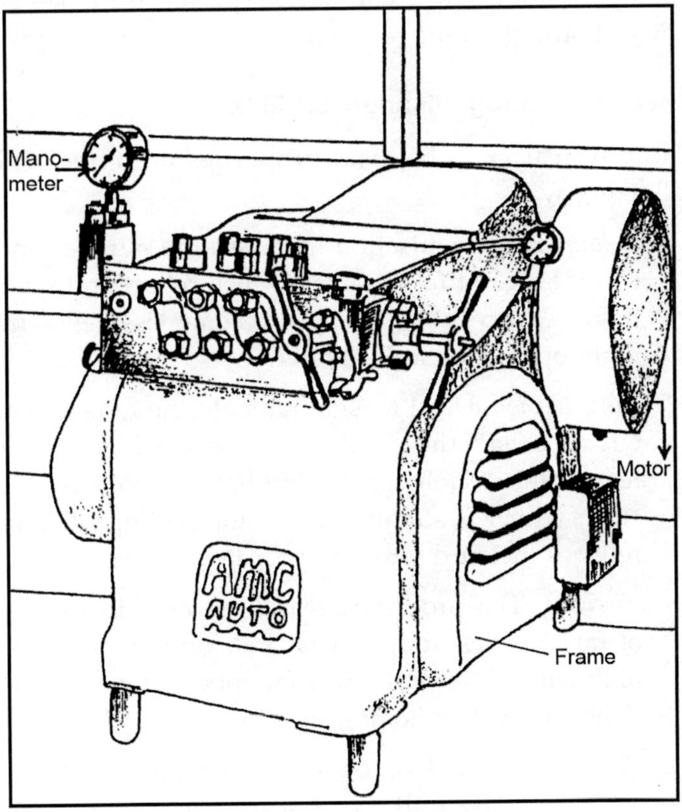

Fig. 7.1: Two stage of homogeniser

Procedure of Manufacturing Homogenised Milk

All homogenised milk must be pasteurised prior to homogenisation or immediately after and the milk should be clarified after homogenisation.

Flow Diagram of Manufacturing Homogenised Milk

Receiving milk—cooling to 5°C and bulk storage—pre-heating (35-40°C)—filtration/clarification——cooling to 5°C——standardising and storage (5°C)—pre-heating (60°C)—homogenisation (2500 psi, 60°C)—pasteurisation— cooling (5°C)—bottling—storage (5°C)

Effect of Homogenisation on Milk

The physical and chemical properties are affected in the following ways:

1. *Colour:* Homogenisation increases the number and total surface area of fat globules, which reflect and scatter the light. As a result cows milk become less yellowing and more opaque as well as white.
2. *Curd tension:* Curd tension refers to the force required to cut or break the curd in atmospheric pressure and temperature. Homogenisation helps in formation of soft curd, which is suitable for infant feeding. This process reduces the curd tension upto 70-73 per cent.
3. *Viscosity:* This process increases the viscosity by clumping of fat globules and increased absorption of casein. The milk with somewhat lower fat appears more viscous and richer milk after homogenisation.
4. *Titrable acidity:* Homogenisation increases the titrable acidity of raw milk but not in case of boiled or over-pasteurised milk. This increased acidity of milk is related with fat content, which is affected by lipase enzyme.

5. *Surface tension:* Surface tension of pasteurised or sterilised milk is more affected than raw milk by homogenisation. It is decreased in homogenised raw milk but increased in heat-treated homogenised milk.
6. *Specific gravity:* Generally, the specific gravity is not affected but in case of large amount of air trapping during homogenisation, it may be slightly decreased.

Factors Affecting Homogenisation

1. *Temperature:* The milk should be pre-heated before homogenisation at about 65-70°C to make fat in liquid state for proper subdivision. The enzyme lipase should be inactivated preferably prior to homogenisation or immediately afterwards.
2. *Pressure:* For liquid product with more than 6 per cent fat, two stages homogenisation is needed to prevent fat clumping, 140 kg/cm^2 (2500 psi) at 1st stage and 35 kg/cm^2 (500 psi) at 2nd stage. 140 to 175 kg/cm^2 pressure is sufficient for milk up to 6 per cent fat in a single stage homogenisation.

Advantages of Homogenisation

1. No formation of cream layer.
2. Fat in milk does not churn due to rough handling.
3. Produces soft curd and butter is easily digested. Hence recommended for infant feeding.
4. More palatable due to brighter appearance and richer flavour.
5. Less susceptible to rancidity.

Disadvantages of Homogenisation

1. Increased cost of production.
2. Fat recovery is difficult.

3. More susceptible to production of sunshine flavour defect.
4. Greater tendency for milk seepage through bottle cap.
5. Greater degree of sedimentation.

DEHYDRATION

Principle of Dehydration

The water in a liquid product like milk is removed to give it a solid form containing 5 per cent or less moisture. Commercial methods of drying are based on supply of heat to the product that evaporates water as vapours.

Objectives of Dehydration

1. To remove moisture so as to reduce weight and volume.
2. To reduce cost of storage and transportation of product.
3. To extend shelf life of milk.
4. To conserve the natural properties of raw milk as far as possible.
5. To provide a product which can be used for many food preparation.

Effects of Dehydration on Milk

This process may affect the milk composition in the following ways:

1. *Milk fat:* Milk fat remains as emulsified or free fat form in dehydrated milk. Free fat makes the dried product greasy and leaves the reconstituted milk with an oily film.

2. *Milk protein:* Heat treatment for dehydration destabilises the protein by inducing denaturation. Spray and drum drying result in destruction of lysine.

3. *Milk sugar:* In freshly dried milk lactose remain in amorphous state, which is hygroscopic in nature. Thus milk powder becomes sticky and form cake. Dried milk product also suffers from colour defect like browning.

4. *Milk salts:* Dehydration results in a low solubility of milk salts. Dried milk may acquire an undesirable flavour due to action of copper and iron on lipid oxidation.
5. *Vitamins:* Dried milk is a very good source of vitamin A and riboflavin.

SYSTEM OF DEHYDRATION/DRYING OF MILK

Drying by Cold Treatment

a. Drying milk by freezing out the water and centrifuging.
b. Drying milk by freezing and sublimation by maintaining a vacuum in the vapourising chamber.

Merits

i. Continuous process.
ii. Moisture content of product may be reduced to extremely low value.
iii. Heat-damaged of milk properties is minimal.

Demerits

i. Complicated.
ii. High operating cost.

Drying by Heat Treatment

a. *Film, roller, or drum drying system*: Generally the concentrated milk is applied in a thin film over the smooth surface of a continuously rotating steam heated metal drum or roller and the dried milk film is continuously scrapped off by a scrapper, located opposite to the point of milk application. Then the milk film is ground to make the powder. This may be atmospheric or vacuum drying. A simple drum or twin drum or double drum may be used for this purpose.

Merits

i. Low operating cost.
ii. Better keeping quality of milk powder.
iii. Suitable for small quantity of milk.

Demerits

i. Low solubility of milk powder.
ii. Cooked flavour in reconstituted milk.

b. *Spray drying system*: Preferably the pre-heated and concentrated milk is atomised to form a spray of very minute droplets, which are then directed to a drying chamber containing a current of hot air. The milk particles leave their moisture and dry to a fine powder, which is removed continuously. Hydraulic pressure jet, pneumatic, or centrifugal disc may be used for atomising milk.

Merits

i. Milk powder is superior in appearance, flavour, solubility.
ii. Economical for large quantity of milk.

Demerits

i. Not economical for large quantity of milk.
ii. Complicated.
iii. Large capital investment.

CHAPTER 8

Preparation of Various Milk Products

CHEESE

Cheese means the product obtained by draining whey after coagulation of milk with harmless milk coagulating agent under influence of desirable bacterial culture. It shall not contain any ingredient that is not found in milk, except coagulating agent used, sodium chloride, and anhydrous calcium chloride. Wax when used for covering outer surface shall not contain anything harmful. The colour used for colouring cheese should not be other than annatt, carotene colour or other permitted colour.

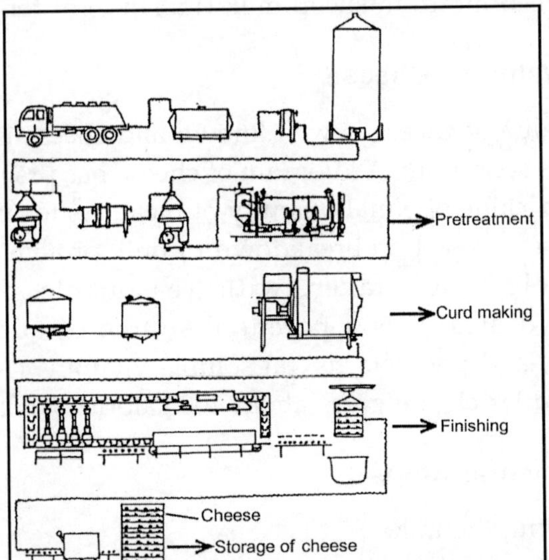

Fig. 8.1: Chees making simplified

Classification of Cheese

Different types of cheeses are based on the following:

1. Original quality of the milk used.
2. Type of coagulation.
3. Degree of processing.
4. Geographical situation.
5. Additives and special culture or process.

Cheese may be named by geographical position of the place of manufacture:

a. Geographical, i.e. manufacturing country or valley:
   Chesire, Lancashire, Derby (England)
   Dorset, Blue, Cantal (France)
   Emmental (Switzerland)
   Roman (Italy)
   Town of selling: Cheddar (England), Camembert (France)
   Institution: Trappist, Port du Salut (France)
b. Quality of milk:
   Zieger–from goat milk (German term for goat milk)
   Pecorino-from sheep milk (Italian term for sheep milk)

Food Value of Cheese

Food value of cheese is well known and cheese is sometimes called as solid milk. A kilogram of cheese has practically twice the food value of equal quantity of meat. As the manufacture of cheese is based on breakdown of milk proteins, it is easily digestible when taken with reasonable quantity of carbohydrates. It is a potential source of milk proteins, calcium, and phosphorus, fat-soluble vitamins (A, D, E, and K). Cheddar cheese gives about 400 calories per 100 g.

Fundamental Steps

1. Setting the milk
2. Cutting or breaking the curd

3. Cooking the curd
4. Draining the whey
5. Salting
6. Pressing the curd
7. Ripening
8. Cutting curd at the right firmness
9. Stirring and heating curd particles to proper consistency
10. Removing cheese from press at optimum elasticity and plasticity.

Some Practical Hints for Cheese Making

1. General precautions: milk should be obtained under best possible conditions.
2. The building and equipment must be suitable.
3. Cheese maker must be trained, experienced, and skilful person.
4. The ripening, after treatment, and other processing should be properly controlled.
5. Test for acidity should be accurately controlled.
6. Starter must be pure and vigorous. It should have a clean acid smell with freedom from gas, spores, and without separation of whey.
7. Rennet should be of best quality.
8. Salt should be pure and best quality with less magnesium.

Composition of Cheese

Table 8.1: Average composition of some important cheese

Ingredient	*Water/moist. per cent*	*Fat per cent*	*Protein per cent*	*Ash per cent*
Cheddar	36.8	33.8	23.7	5.6
Cottage	69.8	1.0	23.3	1.9
Brick	52.53	22.44	20.94	4.81

Contd....

Contd...

Camembert	47.88	26.32	22.21	4.11
Edam	38.07	22.65	30.89	6.19
Emmenthal	33.00	30.50	30.44	4.17
Gouda	54.79	9.02	25.94	5.52
Parmesan	29.63	27.29	34.84	4.76
Roquefort	38.69	32.31	21.39	6.14
Stilton	35.57	31.19	28.96	3.00
Swiss	33.00	30.50	30.40	4.2
Cream	42.70	39.90	14.5	1.9

Source: Outlines of Dairy Technology by Sukumar De (2003)

Requirement of Equipment and Materials (Small Scale Preparation)

i. A rectangular double jacketed cheese vat with gate valve
ii. One vertical and one horizontal cheese knife
iii. Curd rack
iv. Four wooden drying racks
v. Cheese mould
vi. Cheese press
vii. Curd mill
viii. Whey strainer
ix. Salt box
x. Weighing scales
xi. Curd scoop
xii. Dairy floating thermometer
xiii. Large curd knife
xiv. Small buckets
xv. Starter making outfit
xvi. Acidimeter outfit
xvii. Stopwatch
xviii. Rennet testing outfit

xix. Rennet
xx. Cheese colour
xxi. Bandage
xxii. Rennet measuring glass
xxiii. Cheese cloth
xxiv. Needles, thread, and scissors
xxv. Wax
xxvi. Wax colour

Test of Cheese Milk

a. Test for chemical quality:
   i. Fat
   ii. SNF
   iii. Acidity
b. Test for bacteriological quality:
   i. Reduction test
   ii. Coliform test
   iii. Thermoduric test
c. Test for inhibitory substances:
   i. Rennet test
   ii. Curd test.

Preparation of Starter

Starter is generally prepared by autoclaving or sterilising milk. The milk is cooled to 75°F and 2 per cent mother culture is added. The starter is allowed to ripen for 12 hours when the acidity is 0.7 to 0.8 per cent.

Milk sterilised → cooled to 75°F → culture added (2% from mother stock)→ mixed well→incubation temperature maintained for 12 hours or more → curd formed → consistency and acidity checked (0.7-0.8 % desirable).

Cheese manufacture at a glance

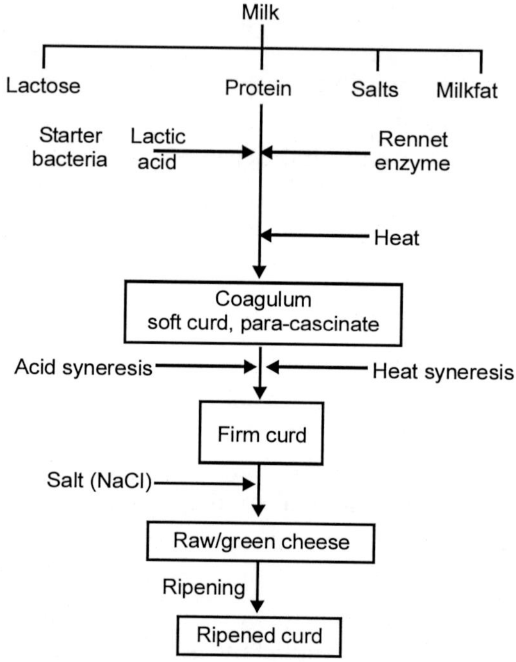

Flow Diagram of Cheese Manufacture

Receiving milk (pasteurised/skim 63°C/ 30 min or 162.5°C/15 sec)

↓

Adding calcium chloride

↓

Adding starter @ 2 per cent (31°C)

↓

Adding rennet (0.13 to 0.15 %)

↓

Curd formed (setting)

↓

Cutting (acidity 0.10 %)

↓

Cooking (upto 37 to 39°C, acidity 0.18 to 0.20 %)

↓

Draining of whey

↓

Cheddaring (acidity 0.25 % temp 100°F)

↓

Salting @ 2 per cent (acidity 0.7 % temp 75°C)

↓

Hopping and vacuum pressing
(27″/50psi, acidity 0.8 %)

↓

Drying

↓

Packaging and storage

Indian Cheeses

Surati Cheese (Panir)

Raw milk→ pasteurised (75°C/15sec) → starter added (20g/100 g at 37°C) → renneting (1 ml/10 kg) → curd formed→ kept in basket → dressed (with 2% salt) → draining of whey → turning → storing (5°C with whey, not more than 48 hrs).

COMPOSITION

- Water 60 to 70 per cent,
- TS 30 to 40 per cent,
- Fat 12 to 20 per cent,
- SNF 18 to 20 per cent.

Dacca Cheese

Raw milk → pasteurised→ starter added→ renneting→ curd→ broken by ladle → basketted in large basket → whey drained by placing a board weighted with stone → cheese removed after 10 days → smoked→ stored.

COMPOSITION

- Water 60 per cent,
- TS 40 per cent,
- Fat 10 to 15 per cent,
- SNF 16 to 20 per cent.

Bandal Cheese

Raw milk (high fat)→ pasteurized → renneting → curd formed → basketted → Dressed with salt → draining of whey → storing with whey.

COMPOSITION

- Water 38 to 42 per cent,
- Fat 35 to 55 per cent,
- Protein 20 to 25 per cent.

Some Common Defects of Cheese

1. Mouldy surface: High humidity during curing and storing, high temperature of curing and storage, poor condition of storage.
2. Corky/hard body: Low fat content, slow acid development, and insufficient moisture before curing.
3. Crumbly body: Excessive acid development in curd.
4. Rubbery body: Low moisture content, low acid development, over salting.
5. Gassy texture/pin holes: Contamination with gas producing organisms.
6. Swiss holes: Contaminated by Propionic bacterium shermanii.
7. Bitter flavour: Poor quality milk, poor quality starter, excessive acid.

Uses of Cheese

1. Direct consumption
2. In sandwiches

3. In preparation of special dishes
4. In the preparation of sauce.

Chhana

Chhana, also called paneer in certain parts of the country is an indigenous whole milk product, which constitutes an important base and filler for the preparation of Bengali milk-sweets. It comprises the milk solids obtained by acid coagulation of boiled hot milk and subsequent straining of whey.

According to the PFA Rules (1976), Chhana is the product obtained from cow or buffalo milk or a combination there of by precipitation with sour milk, lactic acid or citric acid. It should not contain more than 70 per cent moisture, and the milk fat content should not be less than 50.0% of the dry matter.

The acids commonly used are lactic or citric in both natural or chemical form. Although good quality channa necessary for preparation of world famous sweet Rosogolla is invariably made from cow milk, considerable quantity of buffalo milk is also used for its production.

Indigenous Production Methods

Chhana is mostly prepared on a small scale by the disorganised sector which adopts age-old traditional methods and equipment.

Two methods are used for the preparation of Chhana viz, the 'Batch' method and the 'Bulk' method. In common with both methods, all the milk is usually brought to boil togethers in a large iron Karahi.

In the Batch Method

i. The milk which is kept simmering hot, is transferred in small bathes of 0.5 to 1.0 kg to a separate vessel containing already the required amount of cleaned, old sour-chhana.

ii. The mixture stirred with a Wooden ladle.
iii. When clear coagulation has taken place, the Chhana is separated by pouring the contents on a piece of muslin cloth stretched over another vessel
iv. The coagulated milk solids collect on the top of the cloth, while the whey passes through.
v. The cloth containing chhana is then gently removed and kept hung for complete removal of whey and shaping and cooling of chhana whey.

In the Bulk method

i. All the milk (5-15 kg) brought to boil in the Karahi-is then removed from the fire.
ii. The coagulant is then added slowly and gradually in the required quantity to the entire lot of milk
iii. Stirred with the ladle so as to mix it properly and clear coagulation takes place
iv. The chhana is then collected by straining it through a cloth.

Factors Affecting Coagulation of Chhana

i. pH and acidity of coagulation
ii. Temperature and time of coagulation
iii. Strength of the coagulating agent, i.e. acid solution
iv. Type of acid
v. Speed with which the milk is stirred for mixing the coagulating acid.

Conditions for Obtaining a Desirable Body and Texture

i. pH of coagulation should be kept preferably around 5.4
ii. The temperature of coagulation around 82°C
iii. The time of coagulation about 0.5 to 1 minute
iv. An optimum strength of the coagulating acid solution is 1 to 2 per cent

v. Lactic acid gives a granular texture which is suitable for rosogolla making. While citric acid tends to produce a pasty texture suitable for sandesh.

vi. Slow stirring is preferable so as to avoid excessive foam formation

vii. For chhana production, cow milk is preferred because it yeilds a soft-bodied and smooth-textured product.

viii. The chhana from buffalo milk has a slightly hard body and a coarse texture, and not preferred for making good quality chhana sweets

ix. The adulteration of milk with starch results in a gelatinous mass on coagulation, which is not preferred for sweet preparation.

x. The presence of colostrum in milk is also not acceptable because it tends to produce a pasty texture making it unsuitable for sweet-making.

xi. Milk having high initial acidity may coagulate on boiling without addition of any acid, and therefore, not fit for chhana making purpose.

Physical Attributes of Chhana

Types		*Normal physical properties*		
Chhana	Flavour	Colour and appearance	Body and texture	Consistency for sweet-making
Cow	Mild acid	Light yellow, moist surface	Soft and smooth	Good
Buffalo	Mild acid	Whitish; greasy surface	Slightly hard and Coarse	Fair

Source: Outlines of Dairy Technology by Sukumar De (2003)

Yield and Composition

The yield and composition of chhana depends on the initial composition of milk; moisture per cent retained in the chhana and losses of milk solids in the whey.

The average yield of chhana prepared under standard conditions ranges from 16 to 18 per cent from cow milk and 22 to 24 per cent from buffalo milk.

The Composition Chhana

Types of milk	*Moisture %*	*Fat %*	*Protein %*	*Lactos %*	*Ash %*
Cow	53.4	24.8	17.4	2.1	2.1
Buffalo	51.6	29.6	14.4	2.3	2.0

Source: Outlines of Dairy Technology by Sukumar De (2003)

Storage Life of Chhana

i. It depends mainly on the temperature of storage
ii. The average storage life of chhana under ordinary packing at 100± 1°F ± 3°F and 45 ± 5° F is 2,3 and 7 days respectively
iii. Chhana stored at room temperature has a very low keeping quality
iv. Refrigerated storage (5-10°C) is preferable and can be extended to six days
v. The shelf life can further be extended using sodium benzoate, and sodium propionate

Food Value of Chhana

i. Since roughly 90 per cent of the fat and proteins, 50 per cent of minerals and 10 per cent of lactase contained in the original milk is obtained in a fairly concentrated form in chhana, its food value is fairly high.
ii. It is rich in energy-yielding fat and fat-soluble vitamins
iii. It is also rich in muscle building proteins
iv. It is fairly rich in bone forming minerals calcium and phosphorus
v. It is particularly recommended for diabetic patient either salted or cooked with vegetable because of its low content of lactose.

Uses of Chhana

i. It is largely used as a base and filler for the preparation of large variety of Bengali sweets such as Sandesh, Rasogolla, Pantooah, Chhana-murki, etc.
ii. It is also used in the preparation of famous dishes such as Matar-paneer, Sag Paneer, etc.
iii. It is also rich in muscle building proteins
iv. It is fairly rich in bone-forming minerals calcium and phosphorous
v. It is particularly recommended for diabetic patient either salted or cooked with vegetable because of its low content of lactose.

Classification of Indigenous Chhana Based Milk Sweets

Chhana Based:	Rasogolla and its derivatives:
Pure chhana with sugar	A. Rasogolla - Boiled in their sugar syrup
Cooked chhana - juicy products	B. Rossomalai - Boiled in their sugar syrup and transferred to sweet condensed milk C. Rajbhog - Boiled same as Rasogolla, stuffed with Khoa, Spices and cardamom D. Chamcham-Boiled same as rasogolla in cylindri shape, decorated with grated Khoa E. Danadar-Boiled in very thick almost crystalline sugar syrup
Uncooked Chhana	Katchagolla
Cooked Chhana-dry products	Sandesh-medium moisture content Koda pak Sandesh-very low moisture content
Fried Chhana pieces coated with sugar	Chhanar murki
Chhana mixed with coconut mince	Narkel sandesh
Chhana mixed with flour and /Khoa juicy product	Dry product - mixed sandesh Gulabjam, langcha, Kalajam, Chhannar jeelabi

Contd...

Contd...

Chhana dried, drained and smoked and cut into small pieces	Swepi - a product of hill areas

Ref: Outline of Dairy Tech - S De; IDM. 1976 March; IDM - XXXi Dairy Industry Conferance; IDM 54, 10, 2002.

Paneer

Paneer consists of the protein and nearly all the fat, insoluble salts and colloidal materials, together with part of the moisture serum of the original milk in which are contained lactose, whey proteins soluble salts, vitamins and other milk components.

Paneer/Panir may be defined as an indigenous variety of rennet coagulated, soft cheese.

Paneer is primarily acid coagulated milk solids and is extensively used as an ingredient for cooking with vegetables in Northern India.

Cottage cheese prepared by acid coagulation in western countries is a similar product.

The best quality product is made from milk of 6.0 per cent fat content.

It is estimated that 4 to 5 per cent of the total milk produced in India is converted in to paneer. Refrigerated and frozen paneer is sold in several metropolitan cities in India.

The compositional standards of paneer have been defined under PFA Act, 1954 as "Paneer means the product obtained from cow or buffalo milk or a combination there of by precipitation with sour milk, lactic acid or citric acid. It shall not contain more than 70.0 per cent moisture and milk fat content shall not be less than 50.0 per cent of the dry matter. The milk fat content of skim milk paneer shall not exceed 13.0 per cent of the dry matter.

Composition

A. Average composition of fresh paneer made from 6% fat milk:

Fat	25.98%
Moisture	54.67%
Total solids	45.24%

B. Approximate composition of fresh laboratory made surati paneer:

Types of milk	Moisture%	Total solids%	Fat%
Cow	71.2	28.8	13.5
Buffalo	71.1	28.9	13.1

Source: Outlines of Dairy Technology by Sukumar De (2003)

Method of Preparation: Principal Steps Involved

i. *Coagulation of milk:* Buffalo milk having fat content of 6 per cent is taken and heated to 82°C in a cheese vat for 5 min; then cooled to 70°C. The milk is then coagulated in the vat by addition of 1 per cent solution of citric acid. The citric acid is added slowly to the milk while it is continuously stirred, till a clean whey separates out.

ii. *Draining:* When coagulation is complete stirring is stopped and the curd is allowed to settle down for 5 minutes; then the whey is drained out through a muslin cloth and during this time the temperature of whey is not allowed to go below 63°C.

iii. *Hooping and pressing:* The coagulated curd containing all the solids is collected and filled in a hoop, lined with strong and clean muslin cloth. The hoop is rectangular frame with both top and bottom open (measurement 35 × 28 × 10 cm). Both the openings of the hoop have fitted in wooden planks. Pressure is applied on the top of the hoop by placing weight (45 kg) for about 15-20 minutes.

The pressed paneer is then removed from the hoop and cut into 6-8 pieces and immersed in chilled water (4-6°C) for 2-3 hours.

iv. *Wiping and packing:* The chiled paneer is then removed from water and placed on wooden planks for 10-15 minutes. The cut pieces are then wrapped in parchment paper, and stored in a cool place.

v. *Storage:* Paneer can be stored for 6 days at 10°C without much deterioration in quality. The freshness of the product may be lost after 3 days.

Industrial Process for Paneer Making

Flow diagram

Milk
↓
Standardise (5.8% fat or Fat:SNF = 1:1.65)
↓
Heat exchanger (through pump) (heated to 90°C)
↓
Coagulating vat (add acid to pH 5:3 citric acid 1% at 70°C)
↓
Curd
↓
Paneer hoop
↓
Press
↓
Immerse blocks (1-2 hour in coats at 4°C)
↓
Air dry the cool block
↓
Mechanical cutter
↓
Wrapping
↓
Paneer cold storage

Packaging of Paneer

Paneer is highly susceptable to many chemical and microbiological changes. Various packaging materials used by manufactures of paneer include:

i. Co-extended laminates of moisture and oxygen barrier films
ii. Polyethyline sachets
iii. Heat induced shrink film
iv. Wax coated parchment paper for preserving paneer at refrigeration temperature
v. Flexible packaging films like polypropyline (with higher gauges) and
vi. Retort pouches hold a great promise for packaging a long-shelf life of paneer.

Extending Shelf Life of Paneer

i. The addition of sorbic acid may extend the shelf life to five weeks at ambient temp. The sorbic acid content of paneer should range from 0.15-0.30 per cent
ii. Dehydration and deep freezing extends the shelf life of paneer.

Types of Paneer/Paneer Types

i. Conventional Paneer
ii. Low-Fat Paneer
iii. Cow-Milk-Paneer
iv. Recombined and Reconstituted Milk Paneer
v. Filled Paneer
vi. Protein-Enriched Filled Paneer
vii. Vagetable Paneer
viii. Soy Paneer
ix. Groundnut Paneer
x. UF paneer

Continuous Sterilized Paneer Production

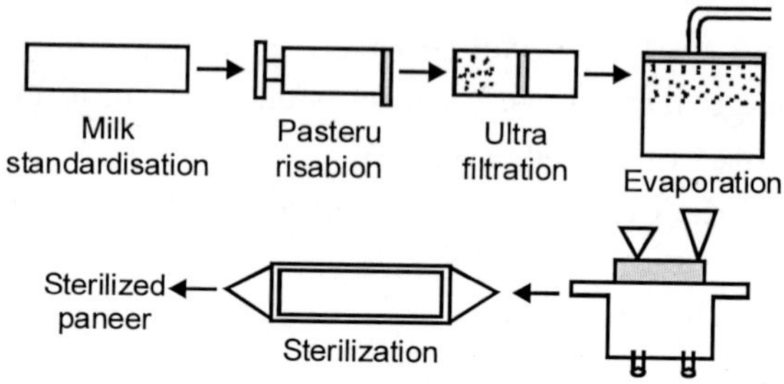

Characteristic Attributes of Paneer

i. Paneer contains approximately 53-55 per cent moisture 23-26 per cent fat, 17-18 per cent protein, 2-2.5 per cent minerals.

ii. Paneer is characterised by typical mild acidic flavour with slightly sweet taste.

iii. It has firm close, cohesive and spongy body and smooth texture.

Composition and yield of various traditional paneer (made from concentrated and ultra filtered milks)

Parameter paneer	*Traditional paneer*			*UF*
	Full fat	*Low fat*	*Skim milk*	
Fat	23.41 (50.84)	8.60 (22.47)	0.20 (0.56)	7.20 (23.51)
Protein	18.33	21.56	25.83	15.92
Lactose	2.40	-	-	5.30
Ash	1.90	-	-	2.21
Total Solids	46.04	38.28	35.42	30.60
Yield	20.00	16.30	14.10	25.00

Ref. IDM P 60; XXXI Dairy Industry Conference

Paneer Curry

A variety of convenient foods have hit the Indian market. Several processes are in vogue for producing this type of foods namely, heat sterilization, drying, frying, baking, fermentation. Therefore, the basic principle of hurdle technology (HT) has been employed to develop shelf stable paneer curry.

Formulation

i. The product consists of gravy and paneer
ii. The water activity of gravy and paneer is reduced to 0.95 by using suitable humectants
iii. pH is lowered to 5.0 by a proper admixing of dahi and skim milk powder
iv. Initially, the gravy is prepared by using onions, tomatoes, using spices and condements, humectants and potassium sorbate (0.1%).
v. Fried paneer cubes and the gravy are packed in tins of retort pouches and subjected to heat treatment
vi. The packages are then cooled in tap water
vii. The retort pouches need to be dipped in scalding water for 5 minutes before consuming the product.

Overall Composition of the Product

Total solids	40.27%
Fat	25%
Protein	5.88%
Ash	3.16%
Carbohydrates and glycerol	6.24%

Shelf Life of the Product

The product has a shelf life of 31 days at 30°C and more than three months at 15°C.

Ref: S De; Rangappa and Achya; IDM 52, 10, 2000; IDM 54, 3, 2002; NDRI Pab No 84.

Butter

As per Federal standard Butter means a food product which is exclusively made from milk or cream or both, with or without common salt and with or without additional colouring matter and containing not less than 80 per cent by weight of milk fat, all tolerances having been allowed for.

According to PFA rules (1976), table or creamery butter is the product obtained from cow or buffalo milk or a combination thereof, or from cream or curd obtained from cow or buffalo milk, with or without the addition of common salt and annatto or carotene as colouring matter.

Specifications: PFA Standard

1. It should be free from other animal fats, wax and mineral oils, vegetable oils, and fats.
2. It should not contain any preservatives except common salt.
3. It should not contain any colouring matter except annatto or carotene.
4. It must contain not less than 80 per cent by weight of milk fat, not more than 1.5 per cent by weight of curd, and not more than 3 per cent by weight of common salt.
5. Total diacetyl (if added) must not exceed 4 ppm.
6. Sodium carbonate/bicarbonate, calcium hydroxide (if added) should not exceed 0.2 per cent by weight of butter.

Bacteriological Standard

1. Coliform count per ml/g (max)-10/g
2. Yeast and mould count (max)-50/g.

Composition of Indian Butter

- Butter fat-80.2 per cent,
- Moisture-16.3 per cent,
- Salt-2.5 per cent and
- Curd-1.0 per cent

Composition of Salted Butter

- Butter fat-80-82 per cent,

- Moisture- 15.6-17.6 per cent,
- Salt-2.5 per cent,
- Protein-1.2-1.5 per cent,
- Lactose-0.6 per cent and
- Milk ash- 0.2 per cent.

Butter is a composite mass containing 80 per cent milk fat and not more than 16 per cent moisture with 2 per cent curd and 2 per cent salt. The butter making process may be considered as a technique to convert small quantities of milk fat present in milk to a composite mass of fats. The changes are physical and change of phase where fat in water emulsion is changed to water in milk fat emulsion.

A good marketable butter depends primarily on the use of pure, clean, and fresh initial product and maintenance of cleanliness in every thing that comes in contact with it.

Principle

If milk is left for some time to stand, a layer of cream appears at the top. The nature of the cream is the same every where except the quantity of fat in it. The fat remains the same as it was in milk but fat globules are only pressed together. If we go on increasing the fat content of cream upto 80 per cent, the product will not be butter but a plastic cream because the fat in cream has still a layer of lecithin and proteins as envelope and remain as emulsion in the moisture. When the envelope is broken, the phase changes from fat in water emulsion to water in fat emulsion and we get butter.

Flow of Process

Milk→ separated to get cream → treatment of cream → conversion to butter→ storage.

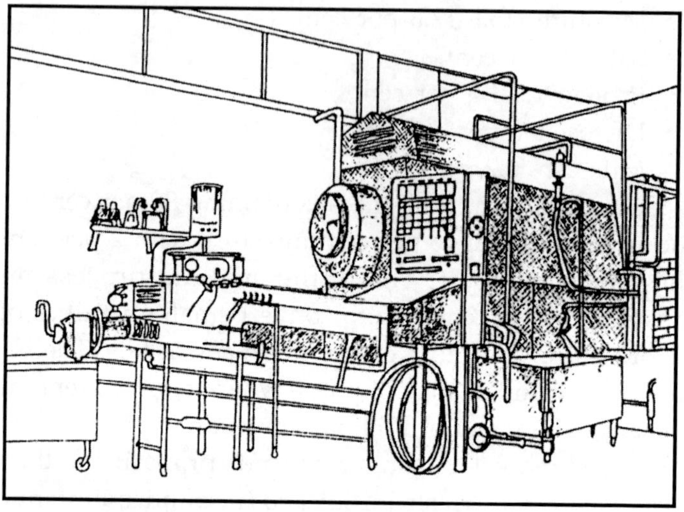

Fig. 8.2: Continuous butter making machine

Manufacturing Flow Sheets

Desi Butter

Milk→ curd→ added with cold water→ beaten by mattris→ added with wash water→ butter.

Creamery Butter

Milk→ separator→ cream→ standardised——————→ salted

standardised ↓ → Butter moisture test; Blending

salted ↓ → Dry salting; Brine salting

Butter

Moisture testblending dry salting brine salting → bulking→ bulk storage→ retail processing→ patting and extruding→ small packages.

Recombination Process

Butter oil→ added with 16 per cent reconstituted milk→ heated to 162.5°F→ homogenised (10.5kg/cm^2) → cooled to 30°C→ added with 2 per cent salt→ mixed well→ cooled to 10°C→ chilled to 4°C→ extruded→ standardised→ bulk package→ storage→ retail package.

Classification of Butter

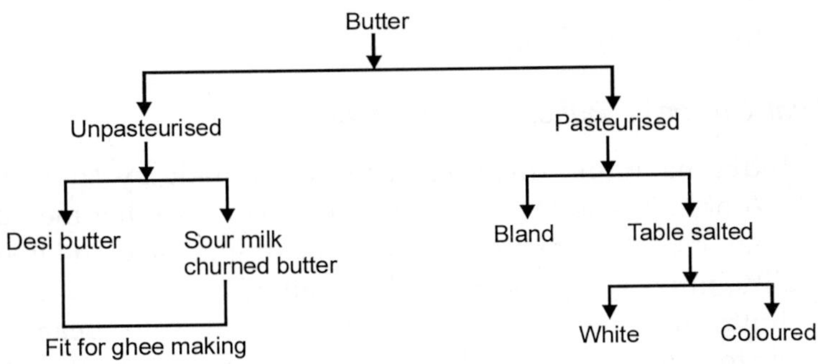

Types of Butter

1. Pasteurised cream butter—obtained from past sweet cream
2. Ripened cream butter—obtained by churning ripened cream, i.e., inoculating the cream with a butter culture
3. Unripened cream butter—made from unripened cream
4. Salted butter—butter to which salt has been added
5. Unsalted butter—butter containing no salt
6. Sweet cream butter—acidity of churned cream is within the limit of 0.20 per cent
7. Dairy butter—made from unpasteurised sour cream
8. Creamery butter—made in a creamery or dairy factory and uniform in quality.
9. Fresh butter—made freshly and not kept in cold storage.

Essentials for Manufacture of High Grade Butter

a. Good quality milk or cream
b. Efficient pasteurisation/vacreation/deodorisation/starter culture
c. Controlled ripening/addition of good dairy flavours
d. Careful and uniform manufacture
e. Modern/clean and sterilised equipment
f. Maintaining proper storage temperature
g. Strict vigil on chemical and bacteriological tests including consumables
h. Attractive packing.

Brief Outline of Butter Making Process

In India maximum quantity of butter is made by churning curd/dahi. This butter is converted to ghee. The other method of making butter is factory method. Maximum yield and good quality butter is produced by this method.

Butter may be made from either sweet or sour cream. In order to produce a desirable flavour the cream is soured by ripening which brought about by encouraging the activities of desired type of microorganisms in the form of culture. After ripening, the cream is churned in a churner to bring about the coalescence of fat globules and to get rid of excess of water and curd in the form of butter milk.

After churning, butter is salted and worked in a butter worker to mix the salt and to get rid of excess water. Fat-soluble colours such as annatto are added during churning to have uniform colour of the butter produced. After working, the butter is made up into packets of different sizes.

Essential Steps in Butter Making

Selection of Cream

Only first grade cream sweet or slightly sour cream will yield good quality butter. Any off flavours present in cream may affect the flavour of the butter.

Neutralisation of Cream

Neutralisation of cream is necessary before pasteurisation if its acidity is high otherwise curd will precipitate during the pasteurisation process. The reasons for standardising acidity of sour cream are:

a. Control of the flavour of fresh butter
b. Control of the amount of fat lost during churning
c. The manufacture of butter of desirable texture and uniform quality
d. Control of the keeping quality of butter.

The acidity of cream should first be determined and the acidity should be reduced to 0.06 to 0.08 per cent before churning for butter to be kept for long storage; cream acidity should be reduced to 0.25 to 0.30 per cent before churning for butter for early consumption.

Meaning of 'Serum Acidity'

The serum acidity refers to the acid content of the fat free portion of the cream or milk. Supposing a sample of cream having a fat content of 30 per cent is tested for acid and is found to contain 0.25 per cent acid, the fat free portion of the cream would be 70 per cent of the sample. If the acid is contained in this part, the acidity of the serum would be:

$$0.25 \times 100/70$$
$$= 0.36 \text{ per cent}$$

If a sample of cream having 40 per cent fat also has an acidity of 0.25 per cent, the acidity of the serum would be:

Per cent serum in sample: 100 –40 = 60
Per cent acid in serum: 0.25 ×100/60 = 0.42 per cent.

Neutralizers Used for Cream

Two types of neutralisers available-the soda compounds and the lime compounds:

Soda Used in Cream Neutralisation

a. Sodium bicarbonate $NaHCO_3$
b. Sodium carbonate-Na_2CO_3
c. A mixture of sodium bicarbonate and sodium carbonate.

Lime Used in Cream Neutralisation

a. Hydrated lime known as calcium hydroxide
b. A mixture of calcium oxide known as quick (CaO) lime and magnesium oxide (MgO).

Pasteurisation of Cream

Reasons for pasteurising cream
a. To kill bacteria, yeasts, and moulds
b. To render enzymes inactive
c. To eliminate off flavours
d. To make butter of uniform quality
e. To make butter of good keeping quality.

Types of Pasteurisation

i. *Vat pasteurisation*—140°F for 30 min or 165°F for 15 min
ii. *Flash pasteurisation*—170 to 205°F for 0.5 to 1 sec
iii. *Plate pasteurisers*—190° to 195°F for 15 sec
iv. *Vacreators*—direct steam injection to 206 to 225°F and use of vacuum 14.7psi or 30″ of Hg.

Vacreation

Vacreation means effectively pasteurising milk, cream or other dairy products, under reduced pressure by steam, followed by a removal of a part of the water vapour as well as volatile substances and finally reducing the temperature of the product quickly in a vacuum chamber to about 110°F. This is nothing but a continuous flow type, high temperature, direct steam, vacuum pasteuriser.

Vacreation is a three-stage process that consists of:

1. Heating the milk, cream mix instantaneously, by mingling with steam under a reduced pressure, in the first chamber to the desired pasteurising temperature.
2. Admitting the product to the second chamber maintained at a lower pressure, where boiling occurs for removal of gases and volatile substances and excess steam to the water condenser.
3. Exposing the product in the third chamber maintained at nearly complete vacuum, where boiling again occurs for additional removal of water vapour and volatile products.

Advantages of Vacreation/ Influences on the Quality of Butter

1. To give higher bacterial destruction
2. To avoid cooked scorched, feed, and weed flavour
3. To improve the keeping quality of butter
4. To result in a finer flavoured butter.

Ripening of Cream

Ripening refers to the fermentation or souring of cream by the addition of 'starters'. The flavours of butter is produced by the fermentation of bacteria in cream. Cultures of *Saturated cremoris* were generally used for cream ripening. The flavour produced is diacetyl by breakdown of AMC. Mixed culture in commercial starters consists of a mixture of *Saturated lactis, Saturated cremoris, Saturated diacetilactis, Leuc. Citrovorum, and Leuc. dextranicum.*

Reasons for Ripening Cream

1. To impart a more desirable flavour to the butter
2. To prevent the development of certain undesirable types of bacteria in butter
3. To make less pronounced certain undesirable flavours such as feed, tallowy, neutraliser, and scorched flavour

4. To aid in making butter of a more uniform quality throughout the year
5. To obtain exhaustive churning, i.e. low fat loss in butter milk.

Churning of Cream

Churning is a process, basically consisting of agitation at a suitable temperature, until the fat globules adhere forming larger and larger masses and until a complete separation of fat and serum occurs.

Reasons for Churning

1. To produce butter
2. To cause foaming and whipping of the cream due to mixing with air
3. To increase the viscosity of cream and to make small grains of fat visible
4. To concentrate and convert that part of milk or cream, i.e. butterfat into a form of food product, i.e. butter
5. To cause reversion of phases changing from the fat-in-water emulsion (cream) to a water-in-fat emulsion (butter).

Theories of Churning

1. The phase-inversion theory by Fischer and Hooker.
2. The foam theory of Rahn.
3. The King theory (combination of phase-inversion and foam theory).

The *phase-inversion theory* states that the churning process involves a reversion of phases, changing from the fat-in-water emulsion (cream) to a water-in-fat emulsion (butter.) The theory based on the belief that the 'membrane' adsorbed to the surface of the fat globules is disrupted by the churning process, causing liberation of some of the fat thereby facilitating fat coalescence and dispersion of butter milk droplets in the free fat. According to this theory the mechanism of butter formation is a progressive coalescence and clumping of the fat globules occurring in geometric progression.

The *foam theory* advanced by Rahn, states that the fat globules gather in the foam caused by the air dispersed in the cream during the churning process. According to this theory, the protein accumulate in the newly produced surfaces, i.e. the 'foam membranes' and surrounds the fat globules, drawn into the foam, lying close together forming clusters. The lumps become larger and disturb the surface equilibrium, the protein in the foam membranes solidifies, the foam collapses, and the butter 'breaks'.

Types of butter churn/ butter making machines:

1. Deshi rotating spindle churn
2. Farm house butter making or hand churn
3. Power churn:
   a. Wooden rotary barrel or box
   b. Internal worker churn
   c. Laminated wooden churn
   d. Metal churns with different shapes, barrel, cylinder, conical, rectangular
   e. Continuous butter making machines.

Earlier forms: Johanssons butter extractor; De laval butter separator (USA); Baltic radiator or Butyrator.

Modern forms: Alfa laval process (Swedish); Cherry-barrel process (USA); Creamery package process; Fritz (Swedish).

Causes of Slow Churning

a. Too low temperature of the cream
b. Too much cream in the churn
c. Viscous cream-excessive richness or lipase action
d. Low testing, watery cream.

Effect of Over Churning

a. The butter granules cannot be washed and chilled properly
b. Composition control becomes difficult
c. The texture and body of the butter become defective (leafy and greasy)
d. The keeping quality of butter is greatly affected.

Washing, Salting, and Working the Butter

Purpose/Reasons for Washing Butter

1. To remove maximum quantity of butter milk.
2. To improve the keeping quality, through the removal of the milk serum contained in the butter milk
3. To control body and texture of butter to some extent
4. To remove certain objectionable flavours
5. To aid in controlling the composition of butter.

Reasons for Salting Butter

1. To prolong the keeping quality of butter
2. To improve the flavour and to have higher relish
3. To aid in preventing the growth of bacteria, yeasts, and moulds
4. To increase overrun.

Amount and Quality of Salt

Usually common salt is added at the rate of 2 to 2.5 per cent of butter fat. The quality of salt should be good, free from any extraneous matter and germs, (specification: IS: 1845, 1961) and low moisture content.

Salting Method

1. Dry salting
2. Wet salting
3. Brine salting

Result of Uneven Salt Distribution

1. Differences in composition of butter
2. Leaky texture
3. Grittiness
4. Waviness in colour

Reasons for Working Butter

1. To mix salt uniformly
2. To expel surplus butter milk
3. To consolidate the granules with a compact mass
4. To incorporate the added make-up water in butter
5. To accomplish a fusion between brine and water in butter.

Overrun

By the term overrun of butter, we generally mean the difference between the kg of butter fat churned and kg of butter made. It may be defined also as the increase in the amount of butter made from a given amount of fat. It is usually expressed as per centage. The overrun is the financial vitamin of the creamery business.

Formula: per cent overrun (OR) = (B – F)/F × 100,

Where B = butter made (kg)

F = fat in churn (kg)

Composition of Butter Milk

Fat-0.10 per cent
SNF- 9.00 per cent
TS- 9.10 per cent
Casein- 2.65 per cent
Albumin- 0.52 per cent
Globulin- 0.21 per cent
Lactose- 4.95 per cent
Ash- 0.74 per cent
Water- 90.9 per cent

Defects of some common creamery butter: (cream and butter)

1. Watery butter- excessive moisture, over churning, washing with water at high temp, insufficient working.
2. Streaky/wavy butter- caused by uneven repining of cream, over-working, uneven salting, bleaching effect of liquids.
3. Greasy butter- want of proper cooling of cream, churning at high temperature, use of fresh cream.
4. Rancid butter- use of rancid cream, copper contamination, lipase activities, fat hydrolysis.
5. Gritty butter- incorrect salting, undissolved coarse salt in butter.
6. Crumbly butter- under-working of butter, sudden chilling of butter immediately after production.
7. Oxidised/ tallowy butter- oxidation of fat due to contact with copper, iron.
8. Leaky butter- under-working of butter, high churning temperature, overchurning of butter.

Uses of Butter

1. Direct consumption with bread
2. As a cooking medium
3. In the baking and confectionery industries
4. In the manufacture of ice cream, butter oil, and ghee
5. In the preparation of sauce.

Food Value of Butter

1. Digestibility value: the digestibility of butter is very high. The coefficient of digestibility of butter is found to be 97.86 per cent
2. Calorific value: the calorific value of butter (80.0 to 85 per cent fat) is 18 calories

3. Vitamins: vitamins that are associated with butter fat and butter are: vit A- essential for normal growth, resistance to infection and disease. Vit B (B1, B2)- necessary for growth, aids digestion, and stimulates appetite, vit B2 commonly known as the anti-pellagra vitamin- essential for growth. Vit C- necessary for normal growth and health, its absence causes scurvy and scorbutic disease, vit D fat soluble, necessary in the diet for bone and skeletal development, helps in retention of calcium and phosphorus.

ICE-CREAM

Ice cream is a frozen dairy food made by freezing a pasteurised mix with agitation to incorporate air and ensure uniformity of consistency. The mix is composed of a combination of milk products, sugar, dextrose, corn syrup in dry or liquid form, water or other optional ingredients with or without eggs or egg products, with harmless flavouring and with or without added stabiliser or emulsifier, and all of wholesome edible material.

According to Federal Standards (USA), it must contain not less than 10 per cent milk fat and 20 per cent total milk solids, except in the case of bulky flavours, the weight of the fat and total milk solids shall not be less than 8 and 16 per cent respectively. It may contain permitted stabilisers and emulsifiers not exceeding 0.5 per cent by weight.

The composition of ice cream varies. The composition for good average ice cream is fat-12 per cent, MSNF-11 per cent, sugar-15 per cent, stabilizer and emulsifier-0.3 per cent and total solids-38.3 per cent. The range in composition may be: fat-8-20 per cent, MSNF-8 to 15 per cent, sugar-13 to 20 per cent, stabiliser-emulsifier-0 to 0.7 per cent and total solids-36 to 43 per cent.

Physical Structure

A three-phase system: Complicated physico-chemical system. Air cells are dispersed in a continuous liquid phase with embedded ice crystals. The liquid phase also contains solidified fat globules, milk proteins, insoluble salts, lactose crystals, stabilisers of colloidal dimension and sugar, and soluble salts in solution. The finished product consists of liquid, air, and solid and thus constitutes a three-phase system.

ISI specifications of ice cream (IS: 2802, 1964)

1. Weight (g/lit) min - 525
2. Total solids (% wt) min - 36.0
3. Milk fat (% wt) min-10.0
4. Acidity (% lactic acid) max – 0.25
5. Sucrose (% wt) max – 15.0
6. Stabiliser/emulsifier (% wt) max 0.5
7. Standard plate count (per g)- not more than 2,50,000
8. Coliform count (per g)- not more than 90
9. Phosphatase test-negative.

Classification

Ice cream and related products are generally classified as frozen desserts which include ice cream, frozen custard, ice milk, sherbet, water ice, frozen confections and mellorine, and parevine type products.

1. Ice cream is a palatable, nutritious, healthful food.
2. Frozen custard is a product similar to ice cream except it contains not less than 1.4 per cent by weight of egg yolk solids.
3. Ice milk is a product that contains not less than 2 per cent or more than 7 per cent fat and not less 11 per cent total milk solids.

4. Sherbet is made of fruits and fruit juices, sugar, stabiliser, and small amount of milk products.
5. Frozen confections include stick novelties.
6. Mellorine type products are similar to ice cream except that the butter fat has been replaced by suitable vegetable or animal fat.
7. Parevine type products are similar to ice cream except that no dairy ingredients are used.
8. Water ices are similar to fruit sherbets except that they contain no milk ingredients.

Classification of Ice-Cream

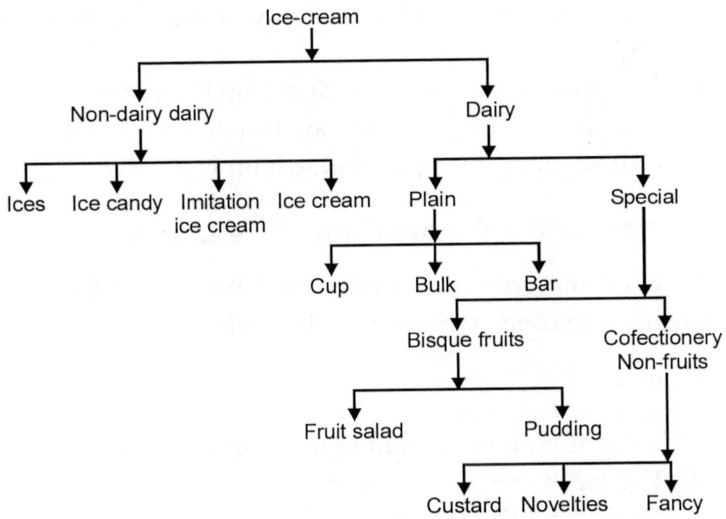

Energy Value and Nutrients in Ice-Cream

The energy value and nutrients in ice cream depend on the food value of the products from which it is made:

1. Ice cream contains 3 to 4 times as much fat and about 12 to 16 per cent more protein than does milk.
2. Ice-cream contains about 4 times as much carbohydrate as milk.

3. The milk solids in ice cream are stored longer before consumption.
4. In addition, ice cream contains other food products such as fruit, nuts, eggs, candies, and sugar that add its nutritive value.
5. Ice-cream is an excellent source of food energy. Since the constituents are almost completely assimilated it makes ice cream an especially desirable food for growing children and for persons who need to put on weight.
6. Ice-cream is a rich source of many essential vitamins like vit A, vit B, vit B_1 (Thiamine) vit B_2 (Riboflavin), and Niacin that are essential for normal growth and good health.
7. The digestibility and palatability of ice cram is also very high.
8. The caloric value of a serving ice-cream (vanilla) containing 12.5 per cent fat, 11 per cent MSNF, 15 per cent sugar and 0.3 per cent gelatin is 211.22.

The Role of the Constituents of Ice-Cream

The basic ingredients are milk fat, MSNF, sweetener solids (sugar), stabilisers, emulsifiers, flavouring, colour, and water.

Milk Fat

1. Milk fat is an ingredient of major importance in ice-cream
2. The fat improves the body
3. Smoothens the texture
4. Constitutes a characteristic richness and mellowiness to ice-cream
5. It is good carrier and synergist for added flavour compounds
6. It promotes desirable tactual qualities
7. It will have a high caloric value
8. An excellent commercial product can be made with 12 per cent butter fat content.

Milk Solids Not Fat (MSNF)

1. MSNF have a water-binding capacity which is of great importance as ice-cream contains 60 per cent water
2. It enhances palatability
3. It increases food value and is also economical
4. It increases viscosity and resistance to melting
5. Improve body and texture
6. Results higher overrun.

Sugar

1. Sweetener for ice cream may be cane sugar (sucrose)
2. Sugar may be used in dry or liquid form
3. Sugar increases the acceptance of the product
4. Makes the product sweeter
5. Enhances the pleasing creamy flavour and desired delicate fruit flavour
6. Sugar in the range of 14 to 16 per cent is most desirable
7. Cheapest source of total solids in the mix.

Stabilisers

1. Used to prevent the formation of objectionable large ice crystals in ice cream
2. Have a high water-holding capacity
3. Effective in smoothening the texture and giving body to the finished product
4. Increases viscosity
5. Prevent coarsening of texture under temperature fluctuations in retail cabinet
6. Gives uniformity of product
7. Improves handling properties.

Emulsifiers

1. To improve uniform whipping quality of the mix

2. Helps to produce drier ice cream with smoother body and texture
3. Produce uniform distribution of air cells in the internal structure of ice cream.

Flavour and Colour

1. Flavour is the most important characteristic of ice cream
2. Have influence on delicacy and palatability
3. Flavour increases the acceptability
4. Colour makes the product attractive and increases its aesthetic appeal.

Outline of making commercial Ice Cream

Add less viscous fluids to more viscous one→ add sugar and gelatin→mix the ingredients thoroughly → pasteurize by heating to 68°C for 30 min→ homogenise the mix (2 stage)→ immediately cool to 4°C→ allow to age (4 to 24hrs)→ frozen (either in batch or continuous freezers)→ packing→hardening and storage (–20 to –30°C).

Home Made Ice-Cream

The common ingredients used for home made ice-cream are milk, sugar, custard powder or corn flour, or ice-cream powder, and desirable flavouring material.

The ingredients are well mixed and heated to boil and then cooled. The cooled mix is poured in the container of a hand-operated freezer. The freezing is done by putting a freezing mixture consisting of salt and chipped ice in proportion of 1:5, round the container. The handle of the freezer is rotated at speed of 100 to 150 revolutions per minute. The temperature of the salt solution is about 21°F. The handle should be rotated continuously until the ice cream has frozen, the ice cream is then allowed to harden in the freezer.

Flow diagram of ice-cream preparation

Selection of ingredients

↓

Calculation for the ingredients of mixes

↓

Preparation of the mix

↓

Pasteurisation of the mix (68°C for 30 min or 155°F for 30 min = batch method or 80°C/175°F for not less than 25 sec= HTST method)

↓

Homogenising the mix (63-77°C, 1st stage 2500 psi, 2nd stage 500 psi)

Cooling the mix (0-5°C/32-40°F)

↓

Aging the mix (5°C/40°F for 3-4 hours)

↓

Freezing the mix (–4 to –3°C/24-26°F for 7 min in batch freezer or –6 to –5°C/21-22°F for 24 sec in continuous freezer)

↓

Packaging of ice cream

↓

Hardening and storage of ice cream (–23 to –29°C/ –10 to –20°F).

Overrun

Overrun is defined as the volume of ice cream obtained in excess of the volume of the mix. It is usually expressed as per cent overrun.

Calculation of Per centage of Overrun

There are two basic methods of calculating per centage over run:

1. By volume
2. By weight

Calculation by Volume

$$\frac{\text{Volume of ice-cream} - \text{volume of mix}}{\text{Volume of mix}} \times 100 = \text{percent overrun}$$

Calculation by Weight

$$\frac{\text{Weight per unit vol of mix} - \text{wt. per unit vol. ice-cream}}{\text{Weight per unit vol of ice cream}} \times 100 = \text{per cent overrun}$$

Factors Determining the Amount of Overrun

1. Legal regulations enforced in the market area
2. Total solids contents of ice cream
3. Bulky flavour ice cream
4. Selling price of ice cream
5. Type of package.

Points to be Considered to Secure Uniform Overrun and Yield

1. Uniformity in refrigerant temperature and rate of flow of refrigerant
2. Use of overrun testers
3. Uniform make etc
4. Not too many freezers per operator
5. Hopper systems for filling containers (in case of batch freezer).

Per centage Overrun for Different Products

	Product	*Overrun per cent*
1.	Ice-cream (packaged)	70-80
2.	Ice-cream (Bulk)	90-100

3. Sherbet	30-40
4. Soft ice-cream	30-50
5. Ice milk	50-80

Some Common Defects in Ice-Cream

1. Acid/sour flavour: high developed acidity, slow cooling of mix, higher temperature of storage.
2. Bitter flavour: Poor quality ingredients.
3. Rancid flavour: Hydrolysis of fat due to lipase activity.
4. Tallowy/oxidised: Fat oxidation due to contact with copper or iron or exposure to sunlight.
5. Crumbly body: Low solids, low stabilisers, and improper homogenisation.
6. Fluppy texture: Low solids, excessive emulsifiers, and excessive overrun.
7. Sandy texture: High MSNF/lactose content, prolong storage.
8. Slow melting: Excessive stabiliser, improper homogenisation.
9. Foamy melt down: Excessive emulsifier, excessive overrun.

GHEE

The word ghee stems from the old Sanskrit 'ghr' meaning bright or to make bright. Ghee is the usual Indian name for clarified butter fat. According to PFA (1976), Ghee, means the pure clarified fat derived solely from milk or from curd or from desi butter or from cream to which no colouring matter or preservative has been added.

Ghee is mostly butter fat free from moisture and curd. Butter contains about 82 per cent butter fat while ghee contains 99 per cent or little more of butter fat. The keeping quality of ghee is more than that of butter. It is widely used in India in preference to butter mostly as a cooking material.

Ghee and its Place in the Indian Diet

i. Ghee is recognised as an important item of Indian diets because of its high nutritive value. Ghee is the only source of animal fat which is predominant in the vegetarian diet in the country.
ii. Ghee is a potential source of energy and yields 9.3 cal/g. According to Hunzikar, ghee is one of the important energy giving foods. It furnishes approximately 3400 cal of heat energy/lb and digestibility is 97.8 per cent.
iii. As a human food, ghee has been considered immensely superior to other fats. Ghee is believed to be a coolant, capable of increasing mental power and physical appearance, and curative of ulcers and eye diseases.
iv. Ghee is an excellent source of fat-soluble vitamins. Vit A and D also provides unsaturated fatty acids which are, but essential for human body metabolism. Vit A is necessary for the growth of children and Vit D plays significant role both in the laying down of calcium and of phosphorous in the bones and in the greater absorption of these elements from the diet.
v. Medicinal qualities of ghee were recognised since long back. The medicinal power of very old samples of Ghee is still widely prevalent and they fetch high value.

Composition of Ghee

Gross Composition of Ghee

i. Bulk of ghee both cow and buffalo milk fat is made up of tri-glycerides (98 %).
ii. The other classes of lipids which are present in minor quantities in ghee are–
   a. Di-glycerides (1-2 %)
   b. Mono-glycerides (0.1-0.2 %)

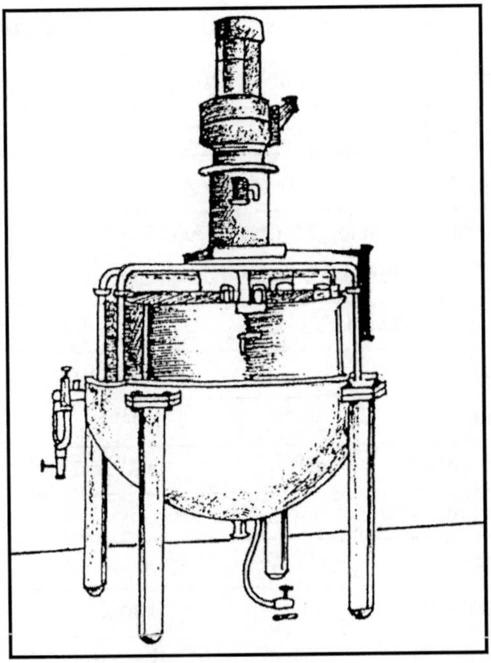

Fig. 8.3: Ghee boiler

c. Free fatty acids (1-10 mg/100 gm)
d. Phospholipids (0-80 mg/100 gm)
e. Sterols, mainly cholesterols
f. Fat soluble vitamins (A, D)
g. Carbonyls (4-6 μg/gm)
h. Glycerols and other (0.8 μM/g)
i. Alcohols (1.8-2.38 μM/g)

Chemical Composition of Ghee

Milk fat – 99-99.5 per cent.
Moisture – not more than 0.5 per cent.
Unsaponifiable matter:

		Cow	*Buffalo*
a.	Carotine	3.2 –7.4 μg/g	-
b.	Vitamin A	19-34 IU/g	17-38 IU/g
c.	Tocopherol	26-48 μg/g	18-37 μg/g

Free fatty acids (% Oleic)- 2.8 (max) (AGMARK).

Table 8.2: Agmark Standard of Ghee

Acidity (% Oleic acid)	*Below 1.4*
Moisture	Not more than 0.3 %
BR reading	40-43
Refractive index	1.4524-1.4545
Baudamin test	(–)ve (N)
RM value	28
Polenske value	1 to 2
Per oxide value	Less than 1.5

Technology of Ghee Making

In principle, ghee making involves three distinct operations:

1. Concentration of milk fat
2. Heat clarification of fat-rich product and
3. Removal of residues produced during the manufacturing of ghee.

According to Srinivasan, the principles of ghee manufacture entail a fermentation process for milk or cream; a mechanical process to gather milk fat in concentrated form; a physico-chemical process of heating at a specified range of temperature to remove moisture and to induce interaction of milk fat with fermented residue of solids-not fat of serum.

The methods of preparation of ghee in India

1. The indigenous method/Desi method.
2. The cream butter process/Creamery butter method.
3. The direct cream boiling method.

In the Desi method, milk is converted to curd and curd churned to desi butter, which can be partially clarified by pre-stratification to yield "kuchha ghee".

Modern dairies use cream separator for concentrating fat up to 40 to 65 per cent level. The cream, with or without fermentation, may be directly heat clarified to ghee (Direct cream method) or converted to butter, which is then made in to ghee (Creamery butter method).

In direct cream boiling method, the cream with or without repining, is to be heated on a uniform low fire with continuous stirring till a temperature of 110 to 120°C is reached. The direct cream method involves the separation of milk in to cream followed by clarification at 110°C for 5 minutes.

Ghee making: The indigenous route

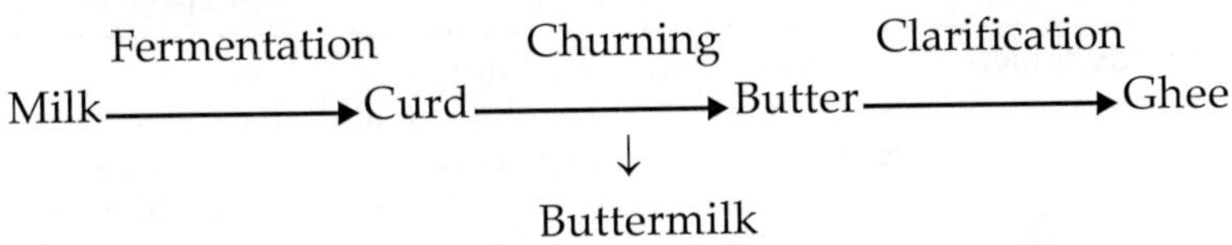

Ghee making: The two-cream routes

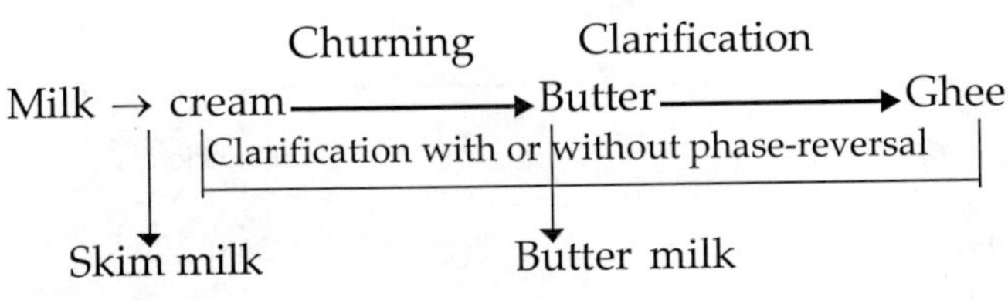

Comparison of the three ghee making processes

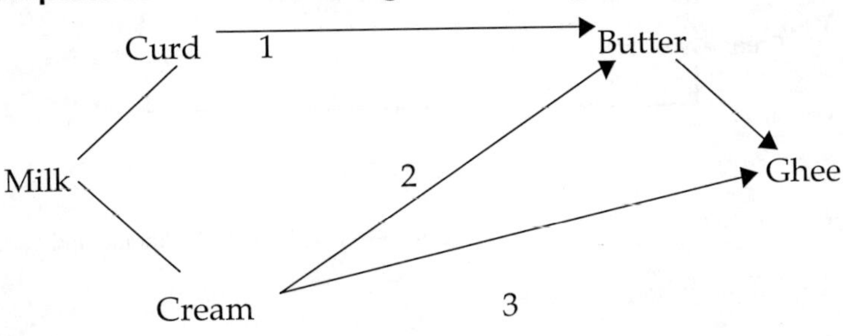

Table 8.3: Comparison of the three ghee making processes:

	I	II	III
Fat recovery, % max	88-90	88-92	92
Aroma	Strong and nuty	Pleasantly rich	Mild
Flavour	Acid	Normal	Flat
Texture	Coarse, grain	Fine grain	Liquid with slight granulation
By-products	Butter milk, ghee residue	Skim milk ghee residue	Skim milk, cream ghee residue
Advantage	Clarification by stratification technique easy and economical	Fat yield high	Max fat yield, keeping quality very high; very rapid process
Disadvantage	Poor yield of fat, poor keeping quality	Medium keeping quality; expensive	Poor sensory characteristics; expensive; slow clarification

Flow diagram of ghee preparation

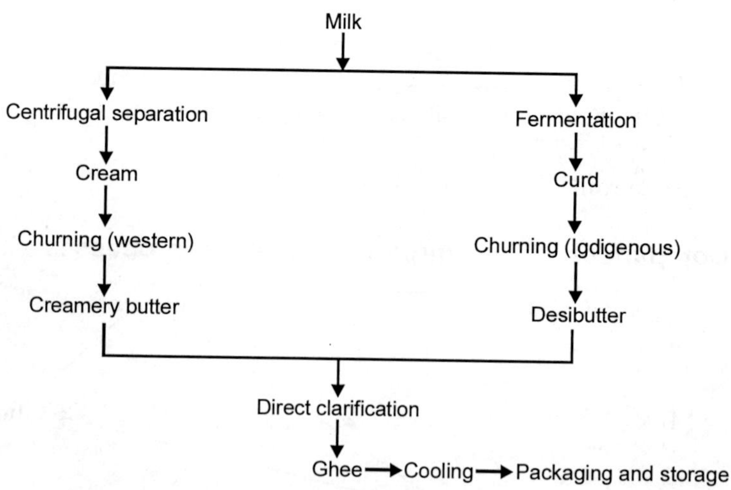

Some Important Physico-chemical Constants of Ghee

1. Melting point→ varies from 28 to 44°C
2. Solidifying point→ varies from 28 to 15°C
3. Specific gravity→ varies from 0.93 to 0.94
4. Refractive index→ varies from 40 to 45
5. Reichert-meissl (RM) value→ not less than 28 (except for ghee from cotton seed feeding areas, where the limit is 20)
6. Polenske (P) value→ not more than 2 (except for cotton seed feeding areas where the limit is 1.5)
7. Saponification value → not less than 220
8. Iodine value → vary from 26-38

Table 8.4: Some characteristics of ghee and fats used as adulterants

	BR index at 40°C	*Baudouin test*	*RM value*	*Polenske value*	*Saponification value*	*Iodine value*
Cow ghee	40-43	– ve	28-33	2.3	225-235	32-35
Buffalo ghee	40-43	– ve	30-36	1.5-2.5	230-237	33-37
Ground nut oil	53-58	– ve	-	-	188-195	84-102
Vanaspati (Hydrogenated)	52-58	+ve	-	-	197-199	70-79
Coconut oil	38-39	– ve	6.8	15-20	245-262	6-10
Palm oil	39-40	– ve	4.8	7-12	243-255	10-18

Table 8.5: Quality standards for ghee

	Standard	*Cu (Max ppm)*	*Fat % Min*	*Moisture % Max*	*FFA % Max*	*Peroxide value Max meq/kg*
1	PFA	30	-	0.5	3	-
2	IS	-	99.5	0.5	0.3	0.8
3	Agmark	-	-	0.3	3	-
4	Codex	0.05	99.3	0.5	-	-

Continuous Ghee Making System

Punjrath (1974) developed the first prototype of continuous ghee making equipment.

Schematic Diagram of Continuous Ghee Making Unit/ System

Raw material (butter) → received in jacketed vat (1) → melting (steam through a pressure reducing valve (V1) → melted butter to the balance tank (pumped by a positive displacement pump (2)] → first stage heat exchanger (5) (through pump (4))→ material heated by steam→ super heated liquid (temperature 112°C) obtained→ flushed in air vapour separator (6)→ fluid (partial moisture removed)→ balance tank (7)→ second stage heat exchanger (9) through pump (8)→ heated to temperature of 112 to 115°C→ vapor separator(10)→ ghee collected in a vat (11)→ clarified in a centrifuge (12).

Advantages

1. Compact design
2. High heat transfer
3. Hygienic operation
4. No or less chance of bulk spoilage
5. Uniform product quality
6. No foaming problems
7. Economy in operation
8. High fat recovery
9. Superior product quality
10. Possibility and ease of in-place-cleaning.

Increasing the Shelf Life of Ghee

The oxidation of unsaturated fatty acids produces hydro peroxides and their subsequent break down products viz. aldehydes, ketones, low molecular weight acids, and, oxyacids, cause the development of off flavours in ghee.

1. The food adulteration rules (1976) allow addition of 0.02 per cent by weight of butylated hydroxy anisole (BHA) and butylated hydroxy toluene (BHT) either singly or in combination in ghee to prevent oxidation and to increase the shelf life of ghee.
2. Betel leaves and curry leaves at 1 per cent concentration has been found to act as replacers for BHA and BHT for extending the shelf life of ghee (135 to 147 days of storage temperature) at 30°C).

Renovation of Ghee/High Acid Ghee

This refers to improve old and rancid ghee so as to make it marketable. This is done by:

1. Re-heating inferior ghee with curd, betel or curry leaves, etc. and filtering it.
2. Adding substance such as saffron, annatto, turmeric juice, etc. to improve the colour of ghee as that of cow ghee.
3. Blending poor quality ghee with a fresh superior quality product.
4. Use of lime of good quality (preferably shell lime, finely powdered) ghee heated to 60 to 70°C powdered shell lime sprinkled on the surface → 3 per cent of ghee temperature quickly raised to 180°C with gentle stirring mass cooled filtered (60°C) improves flavour, granularity, and keeping quality.

Some Common Defects of Ghee

1. Cooked/burnt flavour—This is due to high temperature of clarification of ghee.
2. Rancid flavour—This happen due to lipase action.
3. Oxidised/metallic flavour—Fat oxidation due to direct contact with copper or iron, exposure to sunlight.
4. Greasy texture—It is due to prompt cooling of hot ghee.
5. Burnt colour—It is due to excessively high temperature of clarification.

Uses of Ghee

1. As a cooking or frying medium
2. In confectionary
3. Direct consumption with rice, chapaties, etc.
4. In some indigenous pharmaceutical preparations.

CURD

Curd is a fermented product obtained from pasteurised or boiled milk by souring, natural or otherwise, by a harmless lactic acid or other bacterial culture. It is also known as Dahi. It may be used for churning into desi butter or for direct consumption. For direct consumption three types of curd, viz. sweet, sour, and sweetened curds are available.

Composition of Whole Milk Curd
Water-85 to 88 per cent, fat-5 to 8 per cent, protein-3.2 to 3.4 per cent, lactose-4.6 to 5.2 per cent, ash-0.7 to 0.72 per cent, lactic acid-0.5 to 0.11 per cent.

Curd increases the food and nutritive value as compared to the original milk. It is more palatable, easily digestible than milk and exerts a therapeutic value in stomach and intestinal disorder due to its antibiotic content.

Flow Diagram of Curd Preparation

Milk
↓
Homogenisation
↓
Pasteurisation (85°C for 30 min)
↓
Cooling
↓
Adding starter (0.5-1 per cent)
↓
Incubation (20°C for 16-24 hrs)

↓

Curd

↓

Storage

Str. lactis, Str. diacetilactis, Str. cremoris, are used as starter culture either in single form or in combination with *Leuconostoc spp.*

Sweet dahi has a pleasant flavour, while sour dahi has a sharp and acid flavour. Sweetened dahi has a characteristic brown colour, a cooked and caramelised flavour and firm body because 6.5 per cent cane sugar is added to milk during curd preparation.

LASSI

Lassi, or butter milk is a by-product in the preparation of butter from dahi by the indigenous process. Lassi is a refreshing beverage prepared by stirring up dahi and adding sugar and small quantity of cold water. Lassi is a whitish, viscous, rich aroma, and mild to acidic product.

Lassi-a cultured drink made from dahi is not only a refreshing delicious but also nutritious and easily digestible, due to which it is quite popular amongst all age groups.

Composition of Lassi in Per centage

Water	Total solids	Fat	Protein	Lactose	Ash	Lactic acid
96.2	3.8	0.8	1.3	1.2	0.4	0.44

The factor influencing its composition are-

i. Composition of dahi

ii. Extent of dilution of the curd

iii. Efficiency of fat removal

Some of the physico-chemical and other important attributes of lassi-

i. The titratable acidity of lassi 0.45 to 0.55 per cent.

ii. The pH ranges from 4.01 to 4.56
iii. The viscosity ranges from 18 to 30cp
iv. Viable count ranges from 6×10^8 to 6.7×10^7 cfu/ml
v. Total viable count and lactic acid decrease during storage
vi. Total volatile fatty acids and tyrosine content increase during storage
vii. The titratable acidity and pH increase during storage
viii. Lassi samples heated at 70°C was found to have coarse texture and whey seperation
ix. Lassi samples heated at 55 to 60°C was found to have higher viscosity
x. Addition of 0.2 per cent CMC had good effect.

Flow Diagram of Lassi Preparation

Curd
↓
Churning/broken down/beaten
↓
Addition of water, ice, sugar, flavour
↓
Homogenisation
↓
Filling in bottle/pouches
↓
Cooling
↓
Storage (4°-5°C)

Method of Preparation: (Line diagram)

Raw milk → quality control inspection → filtering → chilling →fore warming → standardisation → boiling for few seconds to 10 minute → cooling to 35 to 40°C → addition of lactic culture (1-2 per cent) → incubation at 35 to 40°C for 5 to 8 hour → breaking the curd → addition of water 20 to 30 per cent →addition of sugar syrup→ pasteurization → cooling → addition of desirable flavour → poly packing → storage in refrigerated condition.

Types of lassi

1. Mildly sour where in mostly Streptococci dominate
2. Highly acidic where in mostly Lactobacilli dominate.

Mechanised Production of Lassi

Significant advancements have been made for the industrial production of lassi by application of UHT process.

In this process milk is first standardized to 9 to 10 per cent SNF and 0.5 to 1.0 per cent milk fat and then fore warmed to 85°C for 30 minutes. Suitable lactic culture (dahi-culture) is added. It is then allowed to ferment at 22°C so as to bring down pH to 4.5. Set curds is beaten up with the help of a stirrer and sugar solution (syrup 30 per cent in water) is added. The product obtained (lassi) is homogenised at 2000 psi and processed at 135 to 145°C for 1 to 5 seconds (UHT), packaged aseptically.

Flow Sheet for mechanised production of lassi

Milk in a multipurpose vat

↓

Standardisation (1.5% fat to 3.84% fat; 9% SNF)

↓

Addition of 0.5 per cent sodium dihydrogen phosphate

↓

Pasteurisation at 90°C/15 minute

↓

Homogenisation (150 Kg/cm^2 1st stage; and 50 Kg/cm^2 2nd stage) at 60°C cool to 31°C

↓

Addition of culture at 31°C

↓

Setting for 12 to 16 hours (pH 4.1)

↓

Breaking the firm curd by agitation

↓

Adding 0.5 per cent low methoxy pectin

↓

Addition of sugar syrup
(12 per cent sugar concentrate in lassi)

↓

Sugar →dissolving in water to make 25 per cent syrup
→straining →pasteurisation → Sugar syrup

↓

Addition of flavour (20 ml/ 100lt lassi)

↓

Homogenization / passing through pump

↓

Mixing well and stand for 1 hour

↓

↓	↓
Pouch filling (non aseptic)	UHT treatment at 110°C/4 second
	↓
	Aseptic packaging
	↓
	Extended life lassi

Importance of lassi in our diet:

i. Lassi improves appetite and vitality.
ii. Controls the adverse effect of drug like Sulphaguanidine on the digestive system.
iii. Its ingestion encourages thiamine synthesis through intestinal micro flora.
iv. Anti bacterial compounds formed and low pH imparted by fermenting flora, help to prevent growth of putrefactive and other undesirable microorganisms.
v. Lassi has been recommended for curing dysentery and other intestinal disorders.
vi. It helps in lowering cholesterol level of blood and minimize the chances of heart attack.

vii. Regular consumption of fermented milk product like lassi helps in controlling the occurrence of cancer.

viii. More easily digestible by persons suffering from lactose tolerance.

ix. Highly nutritive and therapeutic value.

CONCENTRATED MILK PRODUCTS

Kheer

Kheer is a semi solid, viscous milk product. It is an Indian dairy product prepared by boiling and concentrating milk to about one-fourth of its original volume. Kheer also known as Basundi, is a partially dessicated sweetened milk product popular in Gujarat, Maharastra, Karnataka, and Andhra Pradesh.

Basundi is less thickened and normally does not contain flaky /layered texture like Rabri. It is somewhat similar to sweetened condensed milk with the exception that it has pleasant heated flavour and slightly brown colour.

Chemical Composition

Moisture –67.02 per cent
Fat-7.83 per cent
Protein-6.34 per cent
Lactose-8.45 per cent
Ash-1.41 per cent
Sugar-8.95 per cent.

Flow Diagram of Kheer Preparation

Milk
↓
Standardised (4% fat)
↓
Boiling in jacketed kettle (3-4 min)
↓
Constant stirring/scraping

↓

Addition of high grade rice (2.5% of milk)

↓

Gentle heating (3-5 min till final concentration of about 2:1)

↓

Cooling

↓

Packing and storing (refrigerator)

The product is cooled and served chilled. The use of Scraped-Surface-Heat-Exchanger (SSHE) has been recommended for concentration of milk during the manufacture of kheer/basundi.

The nutritive value of kheer is fairly high because it contains all the milk solids in two-fold concentration plus additional sugar.

The out turn of kheer is variable. On an average, the yield of finished kheer is about 50 per cent of the milk used provided the ratio of concentration is 2.

The shelf life of kheer is very limited. The average shelf life is 2 to 3 days at 37±1°C and 10 to 15 days at 4±1°C.

Rabri

Rabri is a partially concentrated and sweetened milk product containing several layers of clotted cream (malai). It is a sweet very popular in northern and eastern parts of the country. Traditionally, it is prepared by simmering whole milk for a long period and adding sugar when the desired concentration is reached.

The Chemical Composition of Rabri

Moisture-30.0 per cent

Fat-20.0

Protein-10.0 per cent

Lactose-17.0 per cent

Ash-3.0 per cent

Sugar-20.0 per cent

Flow Diagram (Traditional method)

Standardised buffalo milk (6.0 % fat)
↓
Simmering in a steam jacket kettle (90°C)
↓
Removal of clotted cream (malai)
↓
Concentration of milk to three-fold
↓
Addition of sugar (@ 5-6 % of milk)
↓
Heat
↓
Cooling
↓
Packing and storing.

Commercial method for manufacture of rabri (using SSHE)

Flow Diagram

Buffalo milk
↓
SSHE
↓
Concentration (2.5 fold)←Sugar @ 6 per cent of milk
↓
Sweetened condensed milk←Shreded Chhana paneer @ 10 per cent of milk.
↓
Rabri
↓
Packing.

Rabri is highly nutritious as it contains all the milk solids about five-fold concentration, plus additional sugar.

The PFA and BIS have fixed no standards for rabri as such wide variation in market samples are found.

The out turn of rabri largely depends on the per centage of total solids in milk, ratio of concentration of milk solids and the amount of sugar added. Buffalo milk produces a more creamy and a higher yield.

Rabri has a very limited shelf life when packaged in polystyrene cups; it has a shelf life of about 18 hrs at 30°C and 20 days at 5°C. The shelf life can be enhanced by use of some permitted preservatives.

Kulfi

Kulfi is an indigenous frozen product. It is the most popular product due to its palatability and nutritive value. It has close similarity with ice-cream and the PFA standards of both are same. The traditional method of preparation of kulfi is very simple.

Milk, while boiling, is sweetened by an addition of sugar and is concentrated to approximately 2:1. It is then cooled and malai, crushed nuts, and flavour are added. The mix is placed in moulds of various shapes and capacities and frozen.

Kulfi is composed of milk or malai with sugar or fruit or lemon juice with sugar.

An increase in total solids to 37 to 40 per cent with 13 to 14 per cent fat and 10 to 12 per cent SNF has been found to improve the body and texture of kulfi.

Though kulfi has resemblance with ice cream, it does not contain air. Its total solid content is about 26 per cent. Concentration of milk to 48 per cent total solids and addition of 3 per cent starch produces best quality kulfi.

Some of the physico-chemical attributes of kulfi are:

i. Surface tension– 50.78 dyne/cm.

ii. pH-6.2

iii. Acidity-0.187

Increase of total solids decrease the pH and surface tension but an increase in acidity.

Flow Diagram of Kulfi Preparation

Milk
↓
Boiling
↓
Addition of sugar
↓
Concentration (2:1)
↓
Cooling
↓
Addition of malai, crushed nut, flavours
↓
Pouring in containers/mould
↓
Frozen (earthen vessel with mixture of ice and salt 1:1)

Khoa

Khoa (Khava, Mava, Palghoa) refers to the milk product prepared by partial dehydration of milk by heating under normal condition.

As per PFA rules (1976), khoa is the product obtained from cow or buffalo (or goat or sheep) milk or a combination there of, by rapid drying. The milk fat content should not be less than 20 per cent of the finished product.

Khoa is thus a partially dehydrated whole milk product where in rapid evaporation of moisture takes place, until the total solid content becomes 70 to 75 %.

Type of Milk Preferred for Khoa-making

Buffalo milk is generally preferred to cow milk for khoa making because buffalo milk gives greater yield besides producing a product with a soft, loose body, and smooth, granular texture desirable for preparation of best quality khoa-based sweets.

A minimum fat level of 4 per cent in cow milk and 5 per cent in buffalo milk is found satisfactory for desirable body and texture in khoa.

Chemical composition:- (per centage)/ min. standard:

Moisture	Fat	Protein	Lactose	Ash	Iron (ppm)	
Cow milk	25.6	25.7	19.2	25.5	3.8	103
Buffalo milk	19.2	37.1	17.8	22.1	3.6	101

(Outlines of dairy technology chemistry (S De)

Bacteriological quality:
Total Count not more than 10,000 /g
Coliform- negative/g
Yeast and mould not more than 100/g

Classification of Khoa

Three main varieties of khoa are prepared, viz. pindi, dhap, and danedar. These varieties vary in quality, texture, and composition, having specific usage.

Table 8.6: BIS specification for khoa

Characteristics parameters	*Requirement for type*		
	Pindi	*Danedar*	*Dhap*
Total solids (% by mass, min)	65	60	55
Fat (% by mass, on dry basis, min)	37	37	37
Total ash (% by mass, on dry (basis, max)	6.0	6.0	6.0
Titratable acidity (as lactic acid-% by mass, max)	0.8	0.9	0.6

Flow Diagram of Khoa Preparation (Traditional method)

Milk in batches taken in karahi/pan

↓

Boiled in open pan over coals/HSDO fired furnace (101°C)

↓

Vigorous agitation or stirring with constantly circular motion (100 rpm or more)

↓

Scraping the sides by a ladle

↓

Removing the pan from the fire

↓

Spreading, working, and scraping with wooden block and ladle

↓

Packing in parchment paper

Principles of Khoa Making

Heating milk in contact with air for a longer period cause heat coagulation of protein. With increased period, the denaturation of protein takes place and the protein, calcium, and phosphate complex are broken. With heat concentration, the fat separates out easily making surface of the product greasy. The caramalisation of sugar takes place causing cooked or burnt flavour.

Out-turn

It depends upon the per centage of total solids in original milk.

$$TS = G/4 + 1.2F + 0.66$$

$$\text{Overrun} = \frac{\text{Khoa} - \text{TS}}{\text{TS}} \times 100$$

Overrun is the per centage increase in weight over total solids of milk as a result of khoa making. This is dependent on the ratio of concentration.

If total solids of milk are 13.31 per cent and total solids of khoa is 75 per cent. The ratio of concentration is 75/13.31= 5.6.

Thus, the average out turn of khoa from 100kg cow milk having 4.5 per cent fat, and 8.55 SNF is 13+13/4= 13+ 3.250= 16.250kg or 16kg 250 gm.

Conditions of Dehydrations

i. *Temperature:* Optimum temp 101°C, higher temp produce coarse texture, burnt-flavour, and dark brown color.
ii. *Speed of stirring:* A speed of 100 to 150 rpm cause good dehydration. Lower speed (30-40 rpm) cause undesirable texture, flavour, and colour.
iii. *Fat level of milk:* A min of 4 to 5 per cent fat is essential, for Gulabjaman, 5 to 8 per cent fat is required.
iv. *Total solids:* More TS yield less Yield: TS of 13.3 to 13.6 is found satisfactory.

Increasing the Shelf Life of Khoa

i. The higher the temperature of storage, the lower is the shelf life.
ii. The product kept in good condition upto 13 days at 30°C, khoa should be filled while hot (70°C).
iii. The shelf life of khoa was increased to 75 days at cold store temp at 7°C.
iv. Incorporation of 0.02 per cent butylated-hydroxy-anisole (BHA) by weight of fat content of khoa increase shelf life.
v. Addition of 0.2 per cent potassium sorbate and mixing with nisin (100μ/gm) has been found to improve the shelf

life of Khoa upto 20 days at 30°C and 125 days at the cold store temp.

vi. Vacuum packaging could enhance the shelf life to 120 days under refrigerated storage.

Physio-chemical Change

i. Gradual reduction in volume of milk till final conversion into solid mass. A change in colour from white to brownish may appear due to partial caramalisation of lactose.
ii. Rapid stirring finely disperses the separated water and lactose present remains as super saturated solution, in the finely dispersed particle of water.
iii. The remaining lactose may crystallise resulting sandy texture.
iv. The main chemical change is the coagulation of milk when the necessary degree of concentration has been reached which is characterised by abrupt changes in colour and consistency.
v. In the first stage, rapid denaturation of albumin and globulin take place and in the second stage, destruction of the protective properties of colloids occurs. This is accelerated by oxidation while aeration takes place during rapid process of stirring.

Defects of Khoa

i. Coarse texture – high temperature.
ii. Crumbly body- neutralised milk.
iii. Greasy body- high fat, low speed of stirring.
iv. Burnt flavour- high temperature.
v. Rubbery body- low fat.
vi. Brown colour- over heating.
vii. Hard body- low speed of stirring
viii. Flat flavour- low temperature.

Improved Systems and Continuous Khoa Making Units

In organised dairies, the karahi and open fire combination has been replaced by a double-jacketed stainless steel kettle with an in-built scrapper. Steam is circulated in the jacket. A village level equipment for khoa making which generates steam in an outer jacket has been developed at NDRI, Karnal.

1. Dodeja et al (1992) developed a continuous khoa making unit using two-Scrapped Surface-heat-exchanger (SSHE). The use of inclined-scrapped-surface-heat-exchanger (ISSHE) developed by Punjrath (1990) has been found satisfactory for continuous production of khoa.
2. The mechanised khoa making system was developed for continuous production of khoa with a rated capacity of 50 lts of milk per hour.

Continuous Khoa Making Plants

1. The first continuous khoa making plant developed by Banerjee et al, which consists of– i. a steam jacketed drum heater, ii. open steam jacketed pan, iii. scrappers, and iv. power drive for scrappers.

   The partially concentrated milk is transferred through a valve to the open pans having reciprocating scrappers. Here the milk is concentrated to a total solid of 70 to 75 per cent. The pilot plant designed and fabricated can convert milk into khoa @ 75 lts of milk per hour.
2. Continuous khoa making plant developed by Dodeja et al (1995) where in two SSHES were arranged in cascade fashion. Milk concentrated to 40 to 45 per centTS→entered the second SSHE (operated in the first SSHE).

At a lower speed of 2.5 rpm— final products collected through the second heat exchanger.

Advantages of Continuous Khoa Making Process

i. Higher heat transfer co-efficients.

ii. Absence of fouling on heat transfer surface.
iii. Pre concentration of the feed in evaporators is not required.
iv. High capacity reduction.
v. Low product inventory.
vi. Hold up value is very small.
vii. Automation system of CIP.
viii. Sanitary operation.
ix. No strain on operators.

Food/Nutritive Value of Khoa

Having practically all the milk constituents intact at the same time being 6 to 7 times concentrated the food value is naturally very high.

1. It contains large quantities of muscle building proteins.
2. It has bone-forming minerals.
3. It supplies energy-giving fat and lactose.
4. It retains most of the fat-soluble vitamins A and D.
5. It also contains water–soluble vitamin-B.
6. The caloric value is 350 to 550 cal/g.

Dehydrated Milk Products

Dehydrated or dried milk products are prepared to reduce bulk for saving space, to reduce cost of packaging and transportation, to improve the storage life of the product, etc. Some of these dried milk products are discussed here briefly.

Butter Milk Powder

Generally butter milk powder is prepared from sweet, sour and high acid butter milk.

Chemical composition of butter milk powder: In case of sweet butter milk, powder contains 2.8 per cent moisture, 5.3 per cent fat, 34.3 per cent protein, 50 per cent lactose and 7.6 per cent ash. In case of sour butter milk, powder consists of 4.8 per cent moisture, 5.7 per cent fat, 37.6 per cent protein, 38.8 per cent lactose, 7.4 per cent ash and 5.7 per cent lactic acid. Both these butter milk powder are high in lactose and minerals.

Flow Diagram of Butter Milk Powders Preparation

Butter milk
↓
Heat (32-49°C)
↓
Fat separation
↓
Heat (85°C for 15 min)
↓
Condensation
(16 per cent total solid in drum process
and 40-45 per cent in spray process)
↓
Heat (71-79°C)
↓
Dry (3-4% water)

Butter milk powder may be used for enriching the nutritive value of prepared food like bread, ice-cream. Sour cream butter milk powder is generally used in animal and poultry feeding and high acid butter milk powder is used for reconstitution into a lassi beverage.

Whey Powder

Whey powder solves several difficulties in relation to production and storage. High lactose content of whey solids creates difficulties in the manufacture of dry whey. Heat causes the lactose to become sticky. Caking from moisture absorption is common particularly during storage.

Chemical composition: in case of cheddar cheese whey, the powder contains 4.5 per cent moisture, 0.9 per cent fat, 12.9 per cent protein, 73.5 per cent lactose, 8 per cent ash and 2.3 per cent lactic acid. The whey powder is fairly high in protein, minerals and very high in lactose.

Flow Diagram of Whey Powder Preparation

Whey
↓
Heat (77-104°C)
↓
Condensation (40-70 % total solid)
↓
Cooling (2-27°C)
↓
Seeding lactose
↓
Spray dry

Whey powder is used as poultry and livestock feed and in the manufacture of processed cheese products, bakery products and frozen desserts.

Cream Powder

Cream powder is a dried milk product, which has a higher fat content (40-70%) than whole milk powder.

Chemical composition: The powder contains 0.8 per cent moisture, 65.0 per cent fat, 13.4 per cent protein, 18.0 per cent lactose, and 2.9 per cent ash.

Flow Diagram of Cream Powder Preparation

Cream
↓
Pasteurisation
↓
Homogenisation (300 psi)
↓
Concentration (40-50 % total solid)
↓
Heat (65-71°C)
↓
Spray dry (1-2 % water)

Fig. 8.4: Spray drier

This is generally used in the confectionary and baking industries, cream cheese, and chocolate, ice-cream, preparation.

Traditional Indigenous Dairy Products of Eastern India

Epilogue or Historical Background

The people of India are traditionally milk conscious and use of milk and ghee has been mentioned in ancient Hindu scriptures like Vedas. Utilisation of milk for conversion into various products like makhan, chhana, dahi, khoa, etc. is prevalent throughout the country for scores of centuries past till date. Queens of Dasaratha were fed with kheer by Rishi for the birth of baby and a full bowl of kheer was offered to lord Buddha by Sujata at the time if self realisation. Lord Krishna is well known as 'Makhan chor'. Chanakya used moth for the destruction of Kush, a wild grass. In fact, no ceremony in India is complete without milk or its products.

Exploring Introduction of Traditional Indigenous Milk Products into Indian Market

As the time passed, the domain of indigenous milk products expanded and products like kheer, khoa, chhana, paneer, srikhand, sandesh, rasogolla, etc. were added up. They form today the backbone of Indian confectionaries. A variety of indigenous milk based products is available in various parts of the country and some of them are region confined and yet to be nationalised.

A glance through the patterns of milk utilisation of the past, reveals a significant scope for setting up small scale dairy units for indigenous milk products. The recommendation of national commission on agriculture (1976) have made it clear that organised dairy industry with the advent of several modern products, should not imply mere denial of traditional milk products to consumers.

Several indigenous dairy products, viz. khoa based; paneer based; curd based; besides ghee and makhan, have been in use under different names, varying from region to region.

National dairy research institute had given due attention to some indigenous products and had come out with such products as cooking paneer, srikhand powder, dried chhana, and dahi powder.

The CFTRI, Mysore also came out with instant dahi powder to make home production of dahi easy and quick.

Attempts were made by NDDB, to manufacture milk sweets on a factory level production. The first product that was taken up, is srikhand under the brand name 'Sugam'.

Mother dairy, Calcutta has successfully launched the first diversification in its product mix 'Sugam misthi dahi'. Such mechanisation in the production of srikhand, gulab jamun, misthi dahi are few notable examples.

Chhana based sweets attracted the acceptability of even western consumers. If the manufacturing of these sweets is standardised and the packaging conditions and the quality standards are improved significantly, the day is not far when chhana based sweets including other milk based sweets will appear in common consumers shops even in western countries.

The development of fermented products in India is preliminary based on the environment, ecology, and taste of the people. This can be categorised on the basis of regions.

- *North zone:* dahi, lassi, desi ghee, makhan.
- *East zone:* payodhi, lal dahi, desi ghee, misthi dahi, gawa ghee.
- *South zone:* dahi, butter milk, desi ghee.
- *West zone:* srikhand, chakka.

Fermented milk, specially lassi, payodhi, misthi dahi have started forming a part of almost every diet in India. Dairy scientists have been given due attention in standardising the technology and manufacture of cultured milks. Carbonated and incorporation of selective additives into the fermented milks have been found successful to enhance shelf life to a greater extent.

The indigenous dairy products and the sweets made from them have great scope in the future dairy industry. The skills possessed by our confectioners (Halwais) for the manufacture of the milk based sweets have no match and it is by virtue of this fact that indigenous milk products making is mostly considered an art rather than a science. This craftsman monopoly technology has now been transformed to industrial level based on source scientific principles.

Table 8.7: Milk based sweets

Type	*Main product*	*Common derived products*	*Places where popular*
Coagulated	Chhana, paneer	Rasogolla, chhana-kheer, sandesh, chhana murki, pantoa, cham-cham	Southern India Northern India
Concentrated partially dehydrated	Khoa/mawa (danedar), dhap /pindi	Barfi, gulab jamun, kalakand, peda (Lal and white)	Northern India , Gujrat
Fermented	Dahi (Lal and sweet) kadhi chakka, srikhand, wadi	Lassi (sweet and salted)	All over India (Gujarat, Maharastra, Karnataka, Madhya Pradesh.
Frozen	Kulfi, malai ka baraf		All over India
High fat	Ghee (kuchha and pucca) makhan and malai sar		All over India
By products	Ghee residue and matha		Northern India

Chhana and paneer: Good quality chhana and paneer can be made from cow milk and buffalo milk respectively. There is a need for enhancing the shelf life of these products at room and low temp; technologies for the mechanised production of chhana and paneer need to be developed. There is also need for the production of low fat paneer in India, since more and more people are becoming aware of the nutritional quality of paneer.

Chhana-based Milk Sweets

Cow/buffalo milk is heated to boiling and held for about 10 mins. When temp lowers down to 70°C sour whey from the previous days batch or dilute citric/lactic acid solution is gradually added with constant but slow stirring till coagulation is complete. Coagulation is allowed to settle and the whey is drained off. The product is creamish white in appearance, compact but spongy body and having sweetish acidic flavour.

Composition

- Moisture-50 to 55 per cent
- Fat-22 to 26 per cent
- Protein-15 to 20 per cent
- Lactose-2.0 to 2.5 per cent
- Ash-1.8 to 2.2 per cent

The recipes for preparation of some typical chhana based sweets on home scale are as follows:

Sandesh

Ingredients:

Chhana-250 gm
Sugar-75 gm
Pista-10 gm
Cardamom (optional)-few nos
Silver paper (optional)-one no

Freshly made chhana is broken into bits and sugar (preferably ground) is mixed into it. It is then put into a karahi and cooked on a very slow fire with constant stirring. Crushed cardamom, if desired is added at the end. When mixture is ready, pour into tray and allowed to cool and set. It is then cut into desired size and shape and decorated, with pista and/ or silver paper.

Rasogolla

Ingredients:
Soft chhana-200 gm
Suji/maida (optional)-8 gm
Sugar-250 gm
Water-1 lt
Elaichi dana-few nos
Flavour (rose)-few drops.

Chhana, maida, suji, and baking powder are mixed to form a dough. Balls are prepared and cooked in boiling sugar syrup during which spongy body and texture develops. The cooked balls are soaked in sugar syrup. Flavours may be added.

Chhana Kheer

Ingredients;
Milk –1 kg
Chhana-150 g
Sugar-40 g
Flavour (optional)-a few drops
Dry nuts (optional)-a few Pieces

Milk is simmered on low heat till reduced to about half the original volume. Sugar is added and simmering is continued. Chhana is kneaded and cut into small cubes, which are added at this stage. Heating discontinued when milk and sugar have penetrated into the cubes.

Chhana Murki

Ingredients:
Chhana –250 g
Sugar-125 g
Water –45 g
Flavour (optional)-A few drops
Dry nuts (optional)-A few pieces

Chhana is kneaded and cut into cubes. Cubes are cooked in boiling sugar syrup till firm. Cooked cubes are then removed from syrup and upon cooling sprinkled with flavour, nuts, and colours.

Chum Chum

Chhana is kneaded into balls and boiled in sugar syrup until desirable body and texture are formed. The surface may be coated with khoa.

Pantoa

Ingredients:
Khoa-300 g
Chhana-300 g
Baking powder-1/2 tsf
Maida –35 g
Sugar –1 kg
Water –1 kg
Ghee –500 g

Chhana, khoa is mixed with maida/suji and baking powder. The mixture is kneaded with water to form dough. Balls are prepared and fried in cooking media till golden brown colour develops. Fried balls are then soaked in sugar syrup.

Khoa Based Sweets

Khoa is concentrated milk in the form of solid mass. It contains 70 to 75 per cent TS, 25 to 30 per cent moisture. Good khoa is white to slightly brown in colour.

Buffalo milk is desiccated with vigorous stirring in shallow pan over direct fire. Gradually a semi solid mass is formed that begins to leave the surface on stirring.

Three types of khoa have been identified.

- *Danedar:* It has granular texture having 20 to 25 per cent fat and 35 to 40 per cent moisture

- *Dhap:* It has fine grain structure and contain 20 to 23 per cent fat and 37 to 44 per cent moisture.

The shelf life of khoa is normally 3 to 5 days at room temperature and 45 days under refrigerated conditions.

Khoa is used as basic ingredient for a large number of sweets like, burfi, peda, gulabjaman, kalakand, etc.

Peda

Ingredients: khoa (mawa)- 225 g
Sugar- 75 g
Pista (optional)- a few pieces
Silver paper (optional)-1 leaf
Cardamom (optional)- a few no

Freshly made khoa broken into bits. Ground sugar is mixed into it and put it into karahi. It is then cooked over a very slow fire (non-smoky) and stirred continuously with khunti. When mixture is ready it is then poured into a tray and allow to cool and set.

Kalakand

Ingredients
Milk- 1 Kg
Sugar- 60 g
Citric acid- 0.5 g
Pista (optional)- a few pieces
Silver paper (optional)-1 leaf
Cardamom (optional)- a few no

Desired quantity of milk is boiled in a karahi over a non-smoky flame with constant stirring in a circular motion. After 10 to 15 minute required amount of citric acid is added with vigorous stirring. When semi solid stage is attained, sugar is added. The product is then allowed to set in a greasy tray and to cool at room temperature.

Burfi

Ingredients of chocolate burfi
Khoa -250 g
Sugar (crystal)- 75 g
Chocolate-10 g

Ingredients of coconut burfi
Khoa-275g
Sugar (crystal)- 135 g
coconut- 35 g

Khoa and sugar (25 to 30 per cent) are mixed in a shallow pan with concomitant heating and stirring. When the body and texture is obtained sthreaded coconut, nuts, and chocolate are added as per desire.

Gulab Jamun

Ingredients
Khoa- 300g
Maida- 35g
Baking powder- 0.5 tsp
Sugar- 1 kg
Water- 1 kg
Ghee- 500 g

Khoa is broken into bits. Baking powder is mixed with maida separately. This mixture is added with broken khoa and thoroughly mixed. The mixture is then kneaded carefully so as to obtain a dough. With such consistency that when made into small it has a smooth un-cooked surface. The balls are then put into sugar syrup, pressing them down in the sugar syrup for sometime so that it soaks in. The gulab jamun is then kept at room temperature for 10 to 12 hours.

Fermented Milk

Fermented milk refers to those milk which have been made by selective use of microorganisms so as to develop the characteristics flavours, body, and texture.

Dahi

Milk is standardised, and pasteurised or boiled. Lactic acid culture is inoculated at 22 to 25°C and incubated for 8 to 10 hours till firm curd is formed and acidity of about 0.8 per cent develops.

The composition of whole milk dahi:

Water- 85 to 88 per cent,

Fat– 5 to 8 per cent,

Protein– 3.2 to 3.4 per cent,

Lactose– 4.6 to 5.2 per cent,

Ash– 0.70 to 0.72 per cent

Flow Diagram

Receiving milk

↓

Pre-heating (35-40°C)

↓

Standardisation

↓

Pre-heating (60°C)

↓

Homogenisation (176 kg /Sq cm)

↓

Pasteurisation (80-90°C/15-30 sec)

↓

Cooling (22-25°C)

↓

Inoculation

↓

Packaging

↓

Incubation (22-25°C/8-12 hours)

↓

Cooling and storage (5°C)

Frozen Milk Products

Kulfi

This is an indigenous ice cream frozen in small containers.

Milk is concentrated in shallow pan. Nuts, flavour, and sugar are added. The mix is then filled in conical metals mould. The product is frozen in a mixture of ice and salty

The composition of kulfi: fat- 7 to 8 per cent; SNF 30 per cent to 37 per cent.

Flow Diagram

Milk
↓
Boiling
↓
Addition of sugar (2:1)
↓
Cooling
↓
Addition of malai, crushed nuts, flavours
↓
Filled in conical/triangular mould
↓
Frozen

Fat Rich Product

Ghee

Traditionally ghee is made by milk-dahi-makkhan-ghee method. In the organised sector, the milk-cream ghee or milk-cream-butter-ghee or milk-cream-butter-pre stratification-ghee methods are employed. The final stage of preparation involves heat clarification/desiccation in steam-jacketed vessel. Ghee residue is an important by-product which could be utilised in a variety of ways for filling of snack foods and for making milk chocolate and other sweets.

The composition of ghee
Fat-99.5 per cent
Moisture- 0.3 per cent

Flow Diagram

Milk
↓
Fermentation
↓
Curd
↓
Churning
↓
Desi butter
↓
Direct clarification (80°C for 30 minute)
↓
Ghee
↓
Packaging and storage

New Product Development

The new product development process requires inputs from various areas. The key features include:

i. Right product to meet the consumer's need.
ii. Blending or modifying the available process technologies with the new.
iii. Managing the process of innovation.
iv. Developing attractive packaging and sound marketing set up.
v. Making manufacturing, the secret weapon of marketing.

Traditional Indian Dairy Products

A great scope exists for Indian dairy industry to exploit the market of the indigenous dairy products by way of adopting mass production technologies. Some of the technologies which can be adopted for industrial mass production are:

1. *Khoa and khoa-based sweets*: With the successful innovation of inclined-scraped surface heat exchanger (ISSHE), these products can be easily adopted by the Indian dairy industry.

2. *Paneer, Chhana, and Chhana-based sweets:* By employing mechanical systems such as caseei process for chhana and paneer making. A continuous chhana making machine designed and fabricated at the NDRI.

3. *Creamed rice:* This is milk-based in-can sterilised product containing a mixture of concentrated whole milk, rice, sugar, flavours, and nuts. Dried fruits may be added optionally. UHT processing is helpful for capturing market.

4. *Cheese-fudge:* This formulation comprises cheese, skimmed milk powder, sugar, corn syrup, cocoa-powder, vegetable fat, and roasted peanuts. Addition of almonds, pistachio, and cashew nuts will improve the test and enrich the product both textual and functional aspects.

5. *Instant dry mixes*: Such products consist of various ingredients mixed in a certain proportion to get desired taste and texture. Amongst these products, gulabjaman occupies a prominent place as a delicacy. Developing instant gulabjaman-mix in dry form from a mixture of non-fat dried milk, semolina, bleached wheat-flour, vegetable fat, and chemical additives is sure to create growing market.

Usage of Dairy Ingredients

Convenience food market is taken as an excellent outlet for the specified dairy ingredients.

Table 8.8: Ready to use milk-solid based dry-mixes

Khoa-based	*Chhana-based*	*Whole-milk based*	*Whey-solid based*	*Acid/Enzyme coagulated*
Khoa powder, kulfi, ice-cream mix powder, Gulabjaman mix powder.	Rosogolla mix powder, sandesh powder.	Chocolate milk powder, coffee complete mango and banana milk shake powder, kheer mix powder	Ready mix for cakes, whey-based fruit drinks in concentrated and dried forms.	Shrikhand-powder, cheese-powder, Lassi powder

The technologies which have exerted on development of new foods consistent with quality are:

1. Development of membrane technology such as ultra filtration, dia-filtration, micro-filtration, reverse osmosis, electro dialysis:
2. Microwave heating for pasteurisation, sterilisation and inactivation of microorganisms.
3. Hurdle technology (e.g. high temperature, low temperature, redox potential, and competitive micro flora)- to prevent/retard microbial growth and to result in stable product at room temp.
4. High-pressure technology – Foods are subjected to high hydrostatic pressure (100-600 Mpa) – to thermally sensitive nutrients, vitamins, colours, flavours, and texture.
5. Modified Atmosphere Packaging (MAP) – to increase shelf life manifold.
6. Intelligent packaging—to protect food from flavour deterioration, discolouration and mould growth.
7. Fractionation of milk fats, cholesterol reduction (stream stripping, sensitive absorption of cyclodextrins, super critical fluid extraction technology).
8. Use of modern centrifugation technologies, e.g. bactofuge, bactocatch, and bactotherm.

9. Fat replacers, e.g. poly dextrose, sucrose polyester, cellulose gel, poly glycerol esters, tri-alkoxycitrate.

Table 8.9: Milk ingredients in convenience foods

Type of Dairy ingredients	*Convenience foods prepared*	*Function*
Cream, butter, butter oil	Canned soups and sauces	Imparts richness and flavour
Skimmed milk powder	Dehydrated soup mixes and frozen soups, dry sauces.	Flavour, imparts richness nutritive value
Sour cream	Pourable salad dressing and gravies	Flavour, creamy appearance
Butter milk solids	Beverage mixes, dehydrated soup mixes	Acidulant, pospholipid rich
Casein	Packaged dry sauces	Emulsifiers, adds viscosity
Sodium caseinate	Dry sauces, dry topping mixes, protein rich textured products	Whitener, imparts richness, emulsifiers, whipping agent
Whey solids	Dehydrated soup mixes, salad dressings, froze pies	Increased total solids, increased nutritive value
Lactose	Dehydrated soup mixes	Mild sweetener
Lactic acid	Salad dressings	Vinegar replacer, preservative
Whey protein concentrate.	Ready mix for cakes, extruded breakfast cereals.	Egg solids replacer, improves nutritional quality.

CHAPTER 9

Packaging and Distribution of Milk

Effective and efficient distribution of consumer articles have always been occupying a key position in a sales organisation of any industrial enterprise. Distribution is largely governed and directed by packaging system and the material used for packaging.

Packaging is the art and the operation involved in the preparation of articles or commodities for carriage, storage, and delivery to the consumers.

Technically packaging is a means of ensuring the safe delivery of a product to the ultimate consumer in sound condition with minimum cost.

Packaging is a sine-qua-non and integral part of food processing which aids in protecting the products during shipment and delivering the products to the ultimate ends (i.e. consumer), in an undamaged conditions. In brief, packaging means placing a commodity into a protective covering or wrapper or container for sound transport and effective storage.

Basic Principles of Packaging

A retail package must sell what it protects as well as protects what it sells. A food package has to perform following primary objectives:

1. It must keep the product clean and provide a barrier against dirt and other contamination.
2. It must protect food from physical damage, moisture, oxygen, and light.
3. The packaging operation must be smooth, efficient, and economic.
4. It should have convenience in usage, release, storage, transportation, and distribution.
5. There should be proper identification, information, and sales appeal.
6. It should protect the product from tampering.
7. It must protect the product from deterioration due to chemical and bacteriological changes.
8. It must facilitate distribution of milk and milk products.
9. It must be attractive to the consumer.

Importance of Packaging

1. Avoid contamination
2. Avoid spoilage
3. Increase shelf life
4. Avoid penetration of external flavour, odour, etc.
5. Creates products appeal or sales appeal.
6. Hygienic in appearance
7. Easily presentable.

Contributing Factors for Design of Packaging

1. Product consideration
2. Market consideration
3. Production consideration
4. Graphics (Product name, symbol, trade name, etc.)
5. Materials for packaging

6. Testing
7. Economics
8. Distribution chain
9. Disposal of used containers
10. Ease and simplicity in handling.

Factors to be Considered for Planning of Good Packaging

1. Water-vapour protection
2. Grease-proofness
3. Infestation
4. Strength
5. Odour
6. Contamination
7. Gas transmission
8. Effect of light.

Properties of the Products to be Studied for Packaging

1. Hygroscopicity
2. Co-hesivity
3. Oxygen-sensitivity
4. Bulk-density
5. Odour-pick up
6. Static electricity
7. Bacteriological status.

Functions of Packaging

A good package must perform three basic functions:

A. To contain the product:
   i. Should be adequately large to hold the product
   ii. Should have proper constructional features so that leakage and spillage do not occur
   iii. Should have enough strength so as to withstand handling, transportation, and storage hazards
   iv. Should be compatible with product.

B. To protect the product:
   i. Should safe guard the product against contamination, degradation due to microbial action, exposure to heat, light, moisture, and oxygen.
   ii. Should safe guard the product from accidental spillage, pilferage, evaporation, etc.

C. To help in promoting the product selling:
   i. Should attract the consumer in its get up
   ii. Should be favourable to dispensation, reclosure, or re-use.

Classification of Milk Packaging Systems

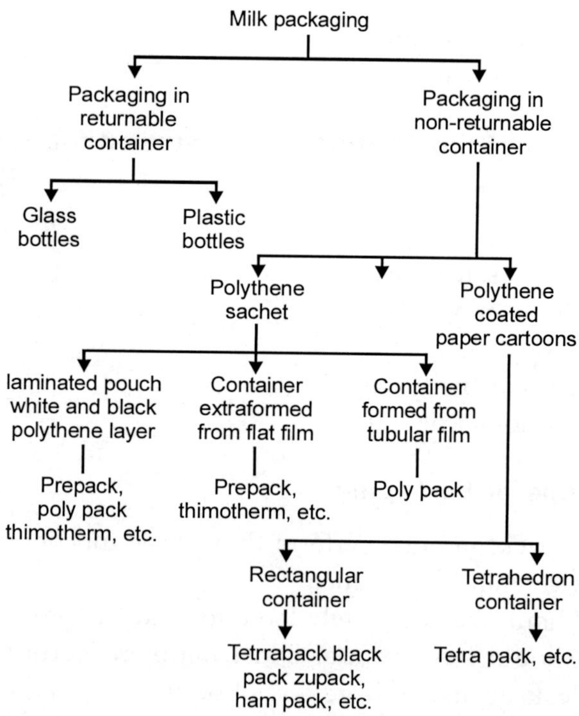

Qualities of Ideal Packaging Material

An ideal packaging material for milk should have following qualities:

1. Should be visually and functionally clean
2. Should be tamper proof
3. Should be transparent
4. Should not have any reaction with milk and milk products
5. Should be strong and durable; preferably light in weight
6. Should be easily disposable
7. Should not absorb any flavour and colour other than milk
8. Should not be affected by heat or detergent
9. Should have sales appeal and attractive
10. Should be cheaper and easily available.

Some Regulatory Attributes of Packaging

The package should be printable to exterior surface and must have following information inscribed on surface:

i. Type of product
ii. Batch number and date
iii. Manufacturers brand name if any
iv. Quantity and quality grade
v. Grade classification there of (ISI/BIS/Agmark, etc)
vi. Maximum retail price (MRP)
vii. Instruction as to how to use and store the product.

Requirements of Packaging Material

i. Should preserve the initial quality of milk for a predetermined length of time
ii. Should not impart smell and taste of milk
iii. Should be resistant to UV light
iv. Should have no physiological effects
v. Should have necessary physico-chemical and mechanical properties

vi. Should lend itself to operation on whatever machines are available.

Packaging Materials and Forms

Different Packaging Material

i. Tin or aluminium containers and boxes
ii. PVC or plastic containers and boxes
iii. Glass bottles or glass containers
iv. Al-foil container
v. Al-foil film
vi. Hard board box or cartoon
vii. Paper cartons
viii. Wax or polythene coated film of paper
ix. Sand witched films with wax, paper, polyfilm
x. Metal film.

Different Forms

i. Bottles
ii. Cans
iii. Cups
iv. Jars
v. Cartons
vi. Sachets
vii. Pouches
viii. Boxes
ix. Tub
x. Sacks
xi. Cask/Barrel
xii. Tumblers

Tin/Tin Plate

i. Consist of tin sheet (0.025 mm thick)
ii. Made up of mild sheet coated on both side with a layer of pure tin

iii. Good strength and excellent barrier properties
iv. Internally lacquered to give greater resistance to corrosion.

Aluminium Foil

i. Common thickness is 0.012 to 0.015 mm
ii. May be coated with lacquer or plastic
iii. Grease proof, shrink proof, non-sorptive
iv. Hygienic, non-toxic, odourless, tasteless
v. Bright in appearance, opaque to light.

Paper Board

i. Kraft paper (bleached/unbleached)
ii. Brown paper
iii. Wax coated paper/plastic coated paper
iv. Paper boards/solid fibre boards/box boards
v. Papers are used in the form of wrapper, carton, bags, cups, and boxes.
vi. Paper boards are used as carton, boxes, etc.
vii. Paper boards include pulp, duplex, triplex, solid, corrugated, moulded.

Plastics

a. Polyethelene
   i. High molecular high density (HMHD)
   ii. High density
   iii. Low density
   iv. Linear low density (LLD)
   v. Shrinkable film.
b. Polypropelene
   i. Co-polymer
   ii. BOPP (Biaxial oriented poly propelene)
c. Polystyrene:
   i. General purpose
   ii. HIPS (High impact polystyrene)

d. Polyester and metalized polyester:
   i. For ordinary film
   ii. For flexible film.

Glass

i. Transparent or opaque
ii. Used in the form of bottles, jurs, jugs, and tumblers
iii. Strength, rigidity, chemical inertness, gas, and water vapour properties
iv. Oldest system of market milk packaging.

Laminates

i. Improves toughness and tear resistance
ii. Improves barrier properties
iii. Improves grease resistance
iv. Effects heat sealing
v. Some typical laminates are
   a. Paper polythene
   b. Cellophane paper
   c. Aluminium foil polythene
   d. Foil polythene
   e. Polyester polythene, etc.
vi. Used as sachet/bag/pouch or carton.

Timber

i. Attractive appearance
ii. Should have necessary mechanical strength
iii. May be treated with casein formalin
iv. May be sprayed with paraffin formalin-wax
v. Used in the form of a box, tub, or barrel.

Carton/Polythene/Wax Coated Paper Carton

i. Made of food grade paper, coated with wax or plastics or of laminates
ii. Maximum space utilisation in vehicles and storage

iii. Tetra pack packaging system uses smallest possible packaging material
iv. Shape and form is most appealing to consumers
v. Protects the product against strains, oxygen, light with good tensile strength, dimensional stability, and rigidity
vi. May be used for liquid, frozen and coagulated milk products.

Flexible Plastic Pouches

i. Milk can be packed in plastic film or laminated pouch
ii. Offer economy, compact storage and easy disposal
iii. Used as wrappers or sachet/bags/pouches
iv. Packaging process can be easily mechanised
v. Protects food from attack by micro-organisms
vi. Only 'food grade' plastic should be used for packaging milk and milk products.

Technological Investigations in Packaging Milk and Milk Products

Milk

The packing material should be water proof, odour proof, preferably leak proof, light proof, have sufficient wet strength.

Butter

Water-vapour proof, odour proof, grease proof, high wet strength storage under 10°C. Aluminium foil laminated with vegetable parchment or grease proof paper and light proof PVC paper.

Processed Cheese

Highly water-vapour proof, high oxygen impermeable, light proof, heat resistant (filling temperature 70°C to 80°C), resistant to weak acids and soluble salts (Lacquered aluminium foil and PVC cups).

Dried Milk Powder

Highly water-vapour proof, odour proof.

Yoghurt

Water proof, leak proof, odour proof (PVC cups).

Dahi

Water proof, leak proof, odour proof (PVC cups).

1. Liquid milk:
   a. Glass bottles
   b. LDPE film in combination with LLDPE or octane based films
   c. Paper laminates-five to six layers film with paper laminates and aluminium foil such as tetra pack for longer shelf life (Dhara/UHT milk).
2. Butter:
   a. Tin container
   b. Aluminium foil/paper board folding cartons
   c. Duplex board with vegetable parchment paper/dry waxed paper/tubes.
3. Milk powder:
   a. Flexible laminates such as BOPP/Al foil/Poly laminates
   b. Tins-nitrogen packed and lacquered from outside/tin cans
   c. Refill packs-laminates with carton, laminated with BOPP/polyester
   d. Bags-lamination used for milk powder and packed in carton/multi wall craft paper bags.
4. Cheese or cheese spread:
   a. Tin containers lacquered from outside
   b. Packed in aluminium foil and in duplex board
   c. Infection moulded polypropylene (PP), high density polyethylene (HDPE) container
   d. Vacuum packed for big blocks.

5. Ghee:
   a. Tin container-inside lacquered
   b. Glass bottle and plastic bottle
   c. HDPE film pouches
6. Chocolate: Aluminum foil wrapped chocolates placed in duplex board cartons.
7. Srikhand/Gulab jamun-Injection moulded/thermoformed plastic containers.
8. Ice-cream:
   a. Thermoformed/injection moulded plastic container.
   b. Duplex board cartons-Poly laminated/polythene coated board.
   c. Laminates/BOPP/polyester used for dolly/candy, etc.
   d. Cylindrical polythene tube with snap on lids.

Table 9.1: Packaging films used for different food/dairy products

Film Type	Products
Polyester-polyethylene/ poly propylene	Meat, sausage, nuts and flips, corn flakes, coffee, tea, butter, ice cream, yoghurt, cheese, milk, custards, sweets and toffees, fish products, and ready cooked foods.
Polyester-aluminium foil-polyethylene/ polypropylene	Ready cooked foods, fruits and vegetable juices, milk shakes, milk powder, cream and dehydrated products, sliced meat and sausages.
Cellophane-al-foil Polyethylene	Spices, instant tea and coffee, vegetable juice, cheese, milk shakes, cream, and dehydrated products.
Bioxially oriented poly propylene (BOPP)	Tea bags, snacks foods, corn flakes, baking and custard powders, confectionary such as toffees, chocolates, lollipops, chewing gums, and milk caramel.
Polyester-high molecular high-density polyethylene	Fat rich products such as butter and ghee.

Aseptic Packaging

Aseptic packaging is the independent sterilisation of product and packaging and then assembling the two in a sterile environment.

The basic object is to:

i. Sterilise the products in a manner that minimises thermal damage to the products and provides higher quality products than would otherwise be available from other processes.
ii. Sterilise the packaging by means that reduce the time temperature of the material.

Aseptic Packaging Adopted by the Food Processing Industry

Aseptic packaging of dairy products and fruit juices in consumer packs has already been extensively used in the markets.

The word 'Aseptic' is becoming well known where all the undesired microorganisms are removed while still retaining the natural qualities of fresh products as far as possible.

In the aseptic installation:

i. The fruit is processed to puree of juice in the normal way.
ii. The remaining oxygen which may cause oxidation and other causes for loss of colour, flavour, and taste, is removed in vacuum.
iii. The oxygen free product is heated in a closed system to 95°C for a short period and rapidly cooled to 20°C to 25°C.
iv. The next step is filling wherein bags with plastic lining; an aluminium middle layer and plastic outer is used.
v. Sterile bags are filled by fully automatic machines after nitrogen has been introduced into the bags to prevent entry of oxygen.

vi. Aseptic packaging in bulk can be done in bags of 5 kg, 20 kg, etc.

While liquid vending machines are used to sell milk and milk beverages.

Solid Vending Machines are of Three Types

i. Column and drawer type
ii. Drop flap machine
iii. Cup-board machine

Liquid Vending Machines

i. Hot or cold liquids or both
ii. Liquid in bulk
iii. Liquid packed in bottle, carton, or sachet.

Packaging Systems in India

There are main two types of packaging systems for filled milk. Depending on the mode of distribution, different types of milk packages are in use. For bulk distribution, rail or road tankers are used, whereas for retail distribution, returnable packaging and single service packaging are used.

Packaging of milk in glass bottles is the oldest system. This system involves collection and transportation of empties to processing plant, washing, sterilisation of dairy bottles, breakage of glass bottles, storage space requirement for empty as well as filled bottles and a long bottling line covering large floor area of the plant. All these factors add the fixed and the variable costs.

Heavy weight and returnability of bottles also has inherent inconvenience both to the distributor as well as consumers.

Non-returnable plastic bottles reduce transport charges through light-weight and one-way delivery but disposal of bottles, in this system, is inconvenient and it is comparatively costlier.

Due to light-weight and one-way delivery polythene sachet has been gaining popularity. It also reduces distribution costs. Packaging material, which forms packs either from a flat film or a tubular film requires very little storage area.

Automatic prepack machine forms a thermo stable film of polythene, makes transverse seal, injects a fixed quantity of milk into the pouch, seals and cuts horizontally and separates the filled pouches. Films are sterilised by passing through U-V light. In our country, flexible and compact type of machines with stainless steel milk extract surface and cleaning in place facilities are commonly used in large dairy plants.

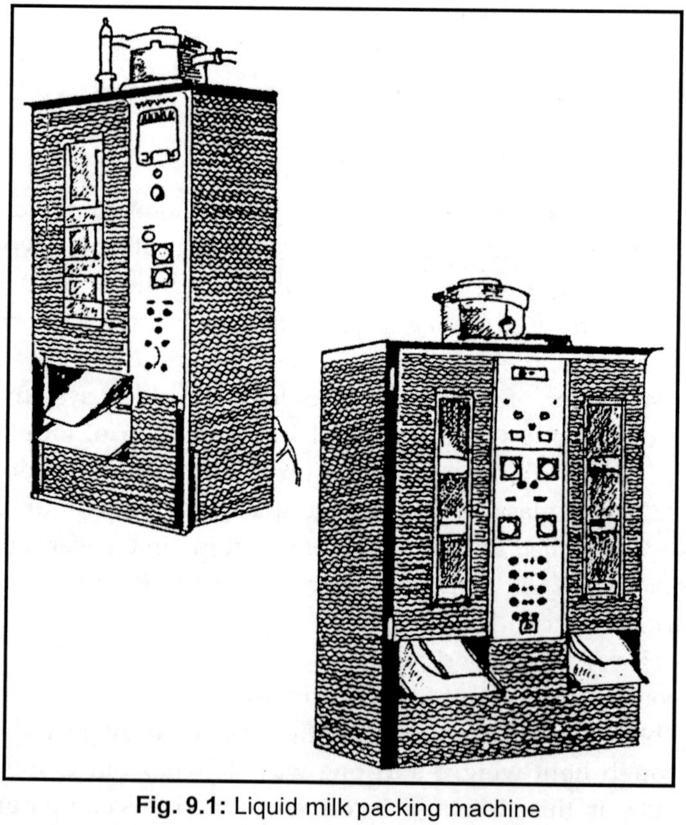

Fig. 9.1: Liquid milk packing machine

Tetra pack packaging system due to tetra hedron shape of carton uses smallest possible packing material. This packaging protects the products against strains, oxygen, light, undesirable aromas, moisture, etc. This system involves lower amount of packaging material, long shelf life at ambient temperature, pilfer proof, and better convenience.

Generally laminated paper consisting of kraft-paper-polyethylene wax or tin aluminum foil-kraft paper polyethylene is used as packaging material. The paper used is an excellent reinforcing medium with good tensile strength, dimensional stability, and rigidity. Shape and form is most appealing to consumers.

Testing of Plastic Films used for the Packaging of Pasteurised Milk

The testing of plastic films in dairy plants is very essential and the quality control of packaging material is very important.

The following test should be performed on the packaging films before accepting for use:

A. Physical attributes:
   i. Thickness of the film
   ii. Width of the film
   iii. Colour of the film
   iv. Weight of the roll of the film
   v. Presence of any holes, scratches, opacity, and conditions of print.

B. Thickness:
   Initially single layer LPDE films of the specifications as below were generally used;
   i. Normal thickness: 95 microns
   ii. Tolerance on thickness: ± 5 per cent
   iii. Normal width: 323 mm
   iv. Tolerance on width: ± 2 mm
   v. Now a days in most of dairy, 300 gauze, i.e. 75 micron thick coextruded and opaque LPDE film is used for packaging pasteurised milk

vi. The transparent film spoils the quality of milk quickly due to entry of light.

For measuring the thickness of film, a thickness gauze meter is used, which indicates the reading directly on a graduated dial. The thickness of the film may be expressed in units such as microns or mm or mil.

100 gauze = 25 micron = 1 mil = 0.025 mm

The thickness of the films can also be found out by the following formula:

$$\text{Film thickness (in gauze)} = \frac{\text{Wt of 100 meters film in g}}{\text{Width of films (cm)} \times 0.46}$$

It is suggested that the thickness of films should always be measured by gauze meter and the test should be done at different places of the film by taking samples at random from different rolls.

C. Heat seal test:

i. Improper heat seal results in great loss to dairy in terms of milk solids and wastage of films.
ii. Improper heat seal also leads to microbial contamination.
iii. The heat seal in a sachet is obtained by sealing test samples of a known width under controlled conditions of temperature, pressure, and dwell time.
iv. The heat seal strength can be tested by peeling apart the sealed strips by using a tensile tester and the value thus obtained can be used for the comparative evaluation of the films.

D. Bursting strength test:

i. This test gives an indication of overall strength of the packaging material.
ii. The principles of the test involves slow and steady increase of the pressure through the elastic membrane either by hydraulic means or by air pressure until rupture occurs.
iii. The equipment for testing is also available.

E. Drop test:

i. The test involves dropping of the filled pouches on a standard drop surface from pre determined heights.

ii. The choice of drop height varies with size and weight of the filled pouch.

iii. The filled one liter pouch of milk should not burst or leak on dropping from 2.5 meters height.

iv. Drop tester are also available.

Method of computing thickness of polyfilm from values of GSM (gm per sq mt).

GSM means gm per sq meter. If a polyfilm is cut in such a size to get a length of one meter and a width of one meter, the weight of the sheet in g represents GSM.

i. Cut length with one meter at its lengthwise and take the weight in g

ii. Multiply by a factor which varies with the width of the film.

For example, for width of 325 mm,

GSM = wt of 1 mt Length × 3.07 (1000 ÷ 325 = 3.07)

Alternatively, cut a film portion measuring 10 cm × 10 cm and weigh the piece. Multiply the weight by 100 to get GSM.

To compute thickness from GSM, use the formula;

$$\text{Thickness} = \frac{\text{GSM}}{0.921} \quad \text{(0.921 is density of polyfilm)}$$

When GSM is 55 g thickness 59.71 m
When GSM is 60 g thickness 65.14 m
When GSM is 70 g thickness 76.0 m

Packaging Procedures Followed in India

Depending upon the mode of distribution, different types of milk packaging are in use. For bulk distribution tankers or milk cans are used whereas for retail distribution, the following types of packaging are done, viz. returnable packaging and single service packaging.

1. *Returnable containers:* individual bottle goes to the consumers and comes back to dairy for reuse, thus known as returnable container. Glass bottles are conventional and popular for milk packaging as returnable containers. Glass bottles have advantages, which are discussed earlier. But the returnable containers have some demerits as follows:
   a. Use of bottles involves bottle washer, which leads to higher cost, floor space, and operational requirements.
   b. Collection of empty bottle increases cost of operation.
   c. Misuse or mishandling of bottles by the workers and the consumers.
   d. Problem of proper accounting of empty bottles returned to the dairy.

ISI has prescribed certain specification for glass milk bottles in IS: 1392, 1983. The following points should be considered during selection of suitable bottles:

1. Shape: The shape is very important from hygienic point of view. It should be such that the cleaning of inner surface does not create any problem to the workers.
2. Capacity: The capacity of bottles should be regular, constant, adopted to the specific filling apparatus, and according to known volume, which a particular plant has decided to offer to its consumers.
3. Resistance: The bottle should be able to withstand temperature and pressure during in bottle pasteurisation or sterilisation.
4. Quality of glass: The bottles should be clear and smooth because irregular composition of glass, air bubbles, or extraneous matter in the glass and deformities not only affect the appearance but also affect mechanical and thermal resistance.

Filling of Bottles

In dairy industry two types of bottle fillers, viz. gravity fillers and vacuum fillers are in use for market milk. Nozzles are arranged in circle in these both types of filling. As the unit comprising supply hopper, filling nozzles and supports for bottle revolves, the bottles are automatically fed into the bottle supports. Then the support raises the bottle so that mouth is pressed against the filling nozzle.

In gravity filler, pressing of the mouth opens the filling valve. The milk flows down the side of bottle and air escapes through hollow valve stem. For uniform filling, the level of milk in hopper should be kept uniform. When the bottle support lowers the filled bottle, the valve closes. These fillers are simple, easily operated but slow in filling and prone to leakage of milk.

The vacuum fillers may be vacuum assisted (single bowl) or straight forward vacuum (double bowl type). In the former type, the typical gravity bowl is closed with airtight cover. For later type, there is a rotary bowl and a float bowl. When the bottle is raised against the filling heads, a seal is formed and air is drawn out through vertical vacuum pipes through the milk pipes through the filling heads and is released into the bottles. At the end of revolution of the filler the lifters on which bottles are carried, are lowered and the seal is broken. This filling is rapid, exact but complicated and cleaning and maintenance is difficult.

Capping of Bottles

The filled bottles are transferred to the revolving capper assembly where bottles are raised in such a way that cap is slided over the bottle and a plunger force the cap into place. Most commonly used methods of bottle sealing are cardboard tops, aluminium caps, and crown corks. Sterilisation of the caps just before use is very essential.

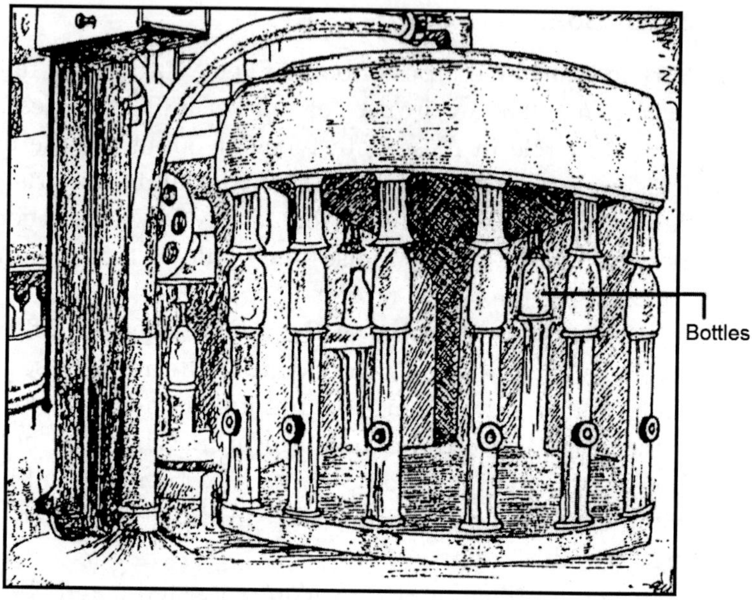

Fig. 9.2: Milk bottle filler in operation

Single Service Container

Two types of single service containers are polyethylene sachet and tetrapak.

Polyethylene Sachet

Automatic prepac machine forms a thermostable film of polyethylene, makes transverse seal, injects a fixed quantity of milk into the pouch, seals and cuts horizontally, and separate the filled bags. Films are sterilised by passing through UV light. In our country flexible and compact type of machines with stainless steel milk contact surface and cleaning–in-place facilities are commonly used in large dairy plant.

Tetrapackaging of Milk

M/s Tetrapak, Sweden, developed an aseptic system for packaging of milk in a disposable container of tetrahedron shape. This system involves lower amount of packaging material, long shelf life at ambient temperature, pilfer proof, better convenience, saving in transportation, etc. Generally, laminated paper consisting of kraft-paper-polyethylene-wax or tin-aluminium foil-kraft paper-polyethylene are used as packaging material. The machine is sterilised by steam or chlorine.

DISTRIBUTION OF MILK

The success of market milk industry largely depends on efficient distribution of milk. In a market milk plant, the last stage of handling is the distribution of milk. Importance of this stage increases day by day because consumers prefer to get a quality product at a reasonably low cost in an attractive package at fixed timing regularly. Milk and milk products are perishable in nature and may get contaminated during distribution, so this process should be rapid and hygienic. Distribution of raw milk and pasteurised milk are discussed below.

1. *Distribution of raw milk:* Raw milk is generally delivered to the consumers under unhygienic condition in poorly designated containers. Raw milk is mostly distributed through conventional types of milk cans made of tin sheets or galvanised iron. These cans are carried either on the shoulders or by cycles. These containers also supply milk to the consumers of cities by rail. The consumers like this door-to-door delivery and convenience in receiving the desired quantity of milk. Sometimes milking is done in consumer's presence in village and town but it is not possible in cities. In this system, the consumers get satisfied the milk is not adulterated and they pay a premium price for the milk.

   Bulk purchaser, viz. sweet makers and local dealers may use aluminium milk can and arrange their own conveyance.

Distribution of raw milk should be discouraged because of lack of sanitation and quality control and fluctuation in milk prices.

2. *Distribution of pasteurised milk:* This process becomes very popular. Where organised dairy plant pasteurises milk, keeps it at low temperature and distributes either in bulk or in retail to individual consumers.

Bulk Distribution

Bulk supplies deliver milk to institution and organisation where a larger volume of milk is required. This process is advantageous because it saves cost but every equipment coming into contact of milk should be sanitised and cleaned thoroughly to maintain strict hygienic condition at every stage during handling.

National dairy development board introduced an easy and flexible electromechanical bulk dispensing of milk under most hygienic and ideal condition, which is known as "Bulk-vending" system. Here, pasteurised milk from plant is brought into storage tank of fiberglass by a road milk tanker. The milk is under refrigeration in storage tank. After insertion of proper coin, a pre-determined quantity of milk is dispensed to the consumer's container. The vending machine or storage tank is cleaned and sanitised (by iodophore) before next filling. The advantages of this system are:

1. Easy to install and maintain
2. Tamper proof
3. Hygienic
4. Easy and flexible operation
5. Quick dispensing
6. Economical
7. All time availability and mass appeal.

Fig. 9.3: Multipurpose processing vat with chiller

Retail Distribution of Pasteurised Milk

Generally glass milk bottles and polyethylene sachets are used for retail distribution of pasteurised milk. The milk is supplied either from a booth or at doorsteps. Sometimes the milk is filled in clean and sterilised metal cans and then cans are resealed from the top. This system is convenient, tamper proof and chance of adulteration can be eliminated in this method. A sales man may carry 2-3 such cans on a tricycle and supply consumers at doorstep. This door-to-door delivery system attracts customers because there is direct contact with salesman. Nowadays most of the dairies are eager to establish their own sales organisation. Milk is delivered to the consumers either directly by the dairy's salesman or through commission agent or depot holder. Sometimes a coupon or advanced monthly cards are issued to desirous customers for regular supply as per their requirements.

Successful Distribution Programme Requires

1. High quality product
2. An attractive package
3. Effective advertising
4. Neat and courteous salesman
5. Pleasing appearance of delivery equipment.

Factors for Effective and Satisfactory Distribution Programme

1. *Proper planning:* Proper planning of sales organisation should have information such as volume of milk to be sold, transport facility, commission agent, advertising, etc.
2. *Route organisation:* Route of organisation should be such that it will be time saving, cost saving, and less person would cover larger area.
3. *Mode of transport:* Depending on volume of milk and distance to be covered trucks, tempo, tricycles, horse cart may be used.
4. *Checking in and out the routes:* Proper record of sold and unsold packs and bottles and regular checking by sales personnel helps in prompt delivery.
5. *Payment and training of route salesmen:* Salesmen may be kept on salary basis or hired on commission basis. Adequate training in selling milk should be given to the salespersons.
6. *Sales outlets:* Sales outlets should be arranged depending on the purchasing power, food habits, availability of milk

through other sources, and other socio-economic consideration.

7. *Cost of milk distribution:* Men and materials should be utilised effectively and efficiently to minimise the cost of distribution.
8. *Frequency of distribution:* In our country milk is to be supplied both in the morning and evening because refrigeration facilities are not adequate.
9. *Daily demand:* The demand for milk fluctuates in different season, special occasion, holidays, etc. so the plant should arrange to tackle the problem.
10. *Attention to customer's complaints:* Consumer's satisfaction should be the main consideration in organising distribution. The complaints must be attended promptly.
11. *Mode of payment by customer:* Consumers pay either in advance or by cash or credit after receiving the supply.
12. *Utilisation of returned milk:* Unsold milk should be tested for quality, which may be deteriorated due to higher temperature, contamination, and tampering by salesperson. Products are prepared if the returned milk is in good condition, otherwise it is used for industrial casein preparation after separation of cream.

CHAPTER 10

Various Standards of Milk and Milk Products

Standards

- It is very much essential to maintain the national and international standards of any products for the acceptance of local or internal consumers or to gain access to foreign markets and sustain the presence of it.
- When the product is "food" the standards of chemical composition of the products and the health hazards, colour, odors, etc. should be considered seriously. Now a days HACCP has come internationally as a major component of standards. The bacteriological quality of milk is vital not only for health point of view but keeping quality or shelf life of the product also depends on it.
- The chemical composition, colour, odor, etc. of raw milk may vary breed to breed, season to season, feed intake, disease condition, etc. Like, raw cow milk of Chandigarh, Haryana and Punjab is having 4 per cent fat whereas raw milk of West Bengal, Bihar, Utter Pradesh, and Assam is having 3.5 per cent. The constituent of raw milk from Jersey and Holstein are having 5.37 per cent and 3.40 per cent fat and 3.92 per cent and 3.22 per cent protein respectively. So for marketing, a standard of those factors

should be maintained where producers and consumers will be benefited.

- Microbial qualities of raw milk also vary depending on the milking process, health status of the cattle, temperature during transportation from collection center to processing center.
- Processing of milk, etc. in tropical countries like us, raw milk will be transported from collecting centers to processing plant at ambient temperature, i.e. 20 to 40°C temperature which is very much helpful for bacterial growth as milk is a very good medium for bacterial growth.

International Standard

Table 10.1: Specific grading requirements for non-fat dry milk

	Extra grade (not Greater than)	*Standard grade (not greater than)*
Milk fat	1.25 per cent	1.5 per cent
Moisture	4 per cent	5 per cent
Titrable acidity	0.15 per cent	0.17 per cent
Solubility index	1.25/ml	2/ml
Bacterial estimate	50,000/g	100,000/g
Scorched particle	Disc B (15 mg)	Disc C (22.5 mg)

Table 10.2: Specific grading requirements for Dry whole milk

	Extra grade	*Standard grade*
Milk fat not less than	26.0 per cent	26.0 per cent
Milk fat not less than	40 per cent	40 per cent
Moisture not more than	4.5 per cent	5 per cent
Bacterial count	50,000/g	100,000/g
Coliform	10/g	10/g
Scorched particles	Disc B (15 mg)	Disc C (22.5 mg)
Solubility index	1 ml	1.5 ml

Table 10.3: Approximate composition and food value of dry milk

Constituents	*Non fat milk*	*Dry whole milk*	*Dry butter milk*
Protein (N × 6.38) per cent	36	26	34
Lactose (milk sugar) per cent		51	38 48
Fat per cent	0.7	26.75	5
Moisture per cent	3	2.25	3
Minerals (Ash) per cent	8.2	6	7.9
Calcium per cent	1.31	0.97	1.3
Phosphorus per cent	1.02	0.75	1.0
Vitamin-A(IU/lb)	165	4950	2300
Riboflavin (mg/lb)	9.2	6.7	14.0
Thiamin (mg/lb)	1.6	1.2	1.2
Niacin (mg/lb)	4.2	3.1	4.5
Niacin equivalent (mg/lb)	42.2	30.6	40.6
Pantothenic acid (mg/lb)	15	13	14
Pyridoxine (mg/lb)	2.0	1.5	2.0
Biotin (mg/lb)	0.2	0.2	0.2
Choline (mg/lb)	500	400	500
Energy (calories /lb)	1630	2260	1700

Table 10.4: All India standards for different classes of milk under PFA rules, 1955 (as amended upto 9th August,1955)

Class of milk	*Designation*	*Minimum per cent*	
		Milk fat	*Milk SNF*
Mixed milk	Raw, pasteurised, boiled, flavoured, and sterilised	4.5	8.5
Standardised milk	Pasteurised, flavoured, and sterilised	4.5	8.5

Contd...

Contd...

Recombined milk	Pasteurised, flavoured, and sterilised	3	8.5
Toned milk	Pasteurised, flavoured, and sterilised	3	8.5
Double toned milk	Pasteurised, flavoured, and sterilised	1.5	9
Skimmed milk	Pasteurised, flavoured, and sterilised	Not more than 0.5	Not more than 8.7

Bacteriological Standards

i. Pasteurised milk (in final container)-plate count/ml should not exceed 30,000

ii. Raw milk

1. Plate count per ml or g

(PC/ml or g)	Grade
a. Not exceeding 200,000	Very good
b. Between 200,000 and 10,00,000	Good
c. Between 10,00,000 and 50,00,000	Fair
d. Over 50,00,000	Poor

2. Direct microscopic clump count

Direct microscopic Clump counts/ ml	Bacteriological quality of milk
Less than 500,000	Good
500,001 to 40,00,000	Fair
40,00,000 to 20000000	Poor
Over 20000000	very poor

International Standards Setting Bodies

1. The American Dry Products Institute (ADPI)
2. The Association of Official Analytical Chemists (AOAC)
3. The International Organisation for Standardisation (ISO)
4. The International Dairy Federation (IDF)
5. The Codex Alimentarius Commission (CAC)

ADPI

i. Earlier known as The American dry milk institute.
ii. Done pioneering work in elaborating standards/grades.
iii. To promote the utilisation of dry milk by making available a uniform grading system.
iv. Serve to protect the manufacturer of superior quality products.
v. Assures uniformity.
vi. Facilitates purchase.
vii. Reduces food safety risks.

AOAC

i. It is association of individual experts from reputed institute, international organisations, and government.
ii. Established in 1884.
iii. The object was to develop precise and accurate standards for analyzing foods, feeds, agricultural chemicals, pharmaceuticals, etc.
iv. The methods are developed through inter-laboratory collaborative studies.
v. Has published a manual containing all the official methods adopted by the association.
vi. This manual contains 1800 test methods.
vii. All the test methods developed by AOAC are voluntary in nature.

ISO (International Organisation for Standardisation)

ISO is the world's largest developer of standards. ISO contribute to make the development and manufacturing and supply of product and services more efficient, safer, and cleaner. This is a network of national standards institutes of 146 countries, on the basis of one member per country, with a central secretariat in Geneva, Switzerland. It is a non-governmental organisation and officially began operation on 23rd February 1947. It published more than 13700 international standards.

ISO 9000 quality systems have 20 elements, viz. management responsibility, quality systems, contract review, design control, document control, purchasing, purchaser supplied product, product identification and testing, trace ability, process control, inspection and testing, inspection, measuring and test equipment, delivery and quality records, international quality audits, training, servicing, statistical techniques. All above 20 elements are applicable for the ISO 9001 model. Some elements are either not applicable or requirement is less rigorous for the ISO 9002 and ISO 9003 models.

Table 10.5: ISO model

Standards and guidelines	*Purpose*
ISO-9000	Provides guidance on the choice of specific model to be adopted for quality assurance in an organisations
ISO-9001	Is meant for manufactures having their own product research and development (R and D)
ISO-9002	Is meant for contract manufactures without any product research and development.
ISO-9003	Is meant for commodity suppliers, having only final product inspection and testing.
ISO-9004:2000	Provides guidance for continual improvement
ISO-10005:1995	Provides guidelines to assist in the preparation, review, acceptance, and revision of quality plans.
ISO-10006:1997	Guidelines to help and ensure the quality of both the project process and the project products.
ISO-10007:1995	Gives an outline to ensure that a complex product continues to function when components are changed individually.
ISO-10012:1997	Provides supplementary guidance on the application of statistical process control when this is appropriate for achieving the objectives of part 1.
ISO-10013:1995	Provides guideline for the development and maintenance of quality manuals, tailored to specific needs.

i. It is a worldwide federation of national standard bodies comprising 120 members, one of each country.
ii. The object of ISO is to promote development of standards and related activities in the world.
iii. The results of ISO technical work are published in international standards.
iv. ISO has elaborated over 10000 international standards.
v. The technical committee on agriculture food products is responsible for the development of ISO standards in the field of food including dairy products.
vi. The body elaborated 55 standards for dairy products which broadly fall under following categories:
    a. Method of test: Physical test, chemical test, and microbiological tests.
    b. Guidelines for sampling of milk products.
    c. Determination of pesticide residues.
    d. Sensory evaluation of food.
vii. All the standards developed by ISO are voluntary in nature.
viii. The agreements for technical barriers to trade (TBT) under the WTO recognises the ISO as one of the international standard setting bodies for parameters such as labeling, packaging, quality agreements, method of analysis, etc.
ix. The ISO standards are particularly important in the field of quality (ISO 9000 series) method of testing and sampling, compositional requirements.

IDF

i. IDF was established in Brussels in 1903.
ii. It is independent, non-political association with the purpose of promoting the solution of scientific technical and economic problems in the international dairy field.

iii. IDF consists of national committee as members, one per country.
iv. It contributes both technically and scientifically in areas, e.g. production, manufacturer, technology, marketing, nutrition, legislation, etc.
v. IDF has developed 180 standards/test method up to January 1998.

CAC

The words codex alimentauius are Latin; meaning food laws or food code.

i. The CAC is an inter governmental body, which come into existence in 1962, under the joint sponsorship of FAO and WHO.
ii. Responsible for formulating international standards for the safety and quality of foods.
iii. At present, 163 countries belonging to FAO and WHO are its members.
iv. Its secretariat is located in FAO Rome.
v. The codex standards define the identity of the product and describe the basic composition and quality factors required for international trade.
vi. CAC has been recognised as the referenced organisation under the SPS (sanitation and phyto-sanitation) agreement of the WTO.
vii. Codex standards are considered as the main international instruments of harmonisation.

BIS

Bureau of Indian standards (BIS), established in 1947 as Indian standard institution (ISI) took up the responsibility of preparing and promoting the general adoption of standards in the country.

The standards are prepared to meet the need of industry protecting the interest of both the producers and consumers.

A list of relevant standards are given below:

IS: 2802-1964	Specification for ice-cream.
IS: 4238-1967	Specification for sterilised milk.
IS: 4709-1968	Specification for flavoured milk.
IS: 4884-1968	Specification for sterilised cream.
IS: 4883-1968	Specification for khoa.
IS: 5162-1969	Specification for chhana.

The standards of weights and measures act, 1976 and the standards of weights and measures (packaged commodities) rules, 1977.

Under the regulations of these act/rules, it is necessary to declare on each package the name and address of the manufacturer, the net quantity contained therein, month/year of manufacture, sale price of the package, etc. As far as possible, this information should appear on the principal display panel and shall be parallel to the base on which the package rests.

Under the above rules, the offences committed against certain requirements are allowed to be compounded at department level by payment of a fine. A written undertaking is also to be given that the particular offence has been committed for the first time and that care shall be taken to prevent future recurrence.

Under these rules, the commodities are to be packaged in specified quantities only:

Commodity

a. Baby food, including weaning food- 200 g, 400 g, 500 g, 1 kg, 2 kg, 5 kg and 10 kg.
b. Butter, cheese, and allied products- 25 g, 50 g, 100 g, 250 g, 400 g, 500 g, 1 kg, 2 kg, 5 kg.
c. Condensed milk- 200 g, 400 g, 1 kg, and multiples of 1 kg.
d. Cooking oils, vanaspati ghee, margarine- 100 g, 200 g, 500 g, 1 kg, 2 kg, 5 kg, and multiples of 5 kg.

e. Ice-creams (bricks)- 50 g, 100 g, 250 g, 500 g, 1 kg, and multiples of 1 kg.
f. Ice-creams (in cups)- 25 g and multiples of 25 g.
g. Milk liquid, sweetened, unsweetened, flavoured- 50 ml, 100 ml, 200 ml, 500 ml, 1 lit, 2 lits, 5 lits and multiples of 5 lits.
h. Milk powder- 100 g, 200 g, 500 g, 1 kg and multiples of 500 g.

CHAPTER 11

Sanitation in Milk Plant

All dairy utensils and equipments after use always contain some milk which after drying form a thin film and encourage bacterial growth. This film of dried milk is difficult to clean and remove. In order to produce milk with a low bacterial count and high keeping quality, it is absolutely necessary to clean and sanitise and dry all utensils.

Cleaning means removal of soil from the surface of dairy equipment. Soil mainly consists of milk and milk product residues which are rich medium for microbial growth.

Sanitisation refers to the process of treating a utensils or equipment with physical and chemical agents in order to destroy most of non-pathogenic and all pathogenic microorganisms. So cleaning and sanitisation should be done simultaneously to make the surface of equipment free from milk residues and viable organisms.

Sanitary requirement for handling or processing or controlling or manufacturing milk and milk products based on the MMPO-1992 (Milk and Milk Products Order)

1. The premises shall be clean, adequately lighted, ventilated, properly whitewashed or painted.
2. There shall be adequate arrangements for disinfections and deodouring.

3. The ventilation and lighting of the building shall be in accordance with the provisions made by or under the Factories Act 1948.
4. The establishment shall be so maintained as to permit hygienic production and all operations including processing and manufacturing milk and milk products shall be carried out carefully under strict hygienic and sanitary conditions.
5. There shall be adequate supply of pure and potable water, free from pathogenical microorganisms.
6. There shall be an efficient drainage system and provision for treatment of refuse and effluents before disposal. Such facilities shall conform to the requirements laid down by the state pollution control board.
7. It should be ensured that employees wash their hands with soap and detergent before they resume work.
8. No person suffering from infections or contagious diseases shall be allowed to work and arrangements shall be made to get the staff medically examined once in six months.
9. The staff working in processing and product manufacturing unit shall be provided with white aprons or uniforms and head gears face masks which shall be clean.
10. Adequate CIP systems shall be maintained.
11. The equipments used for handling milk and milk products shall conform to sanitary standards as may be fixed from time to time by MMPAB.
12. Adequate measures shall be taken to prevent mould growth, infestation of cockroaches and other pests.

Detergent

Detergents are chemical compounds which are used for cleaning and sterilisation purposes. The detergent action is called detergency. Detergency in dairy industry has become a specialised branch of cleaning.

There are two objects of detergency:

a. Primery object: Removal of milk residue and other substances which provide food for bacterial growth and toxicity.

b. Secondary Object: Killing of microorganisms, sterilising and efficiency of cleaning.

Modern dairy detergents are based on following action:

i. Saponification
ii. Wetting and surface activity
iii. Emulsification
iv. Rinsing
v. Buffering
vi. Sterilising power
vii. Chemical disinfections

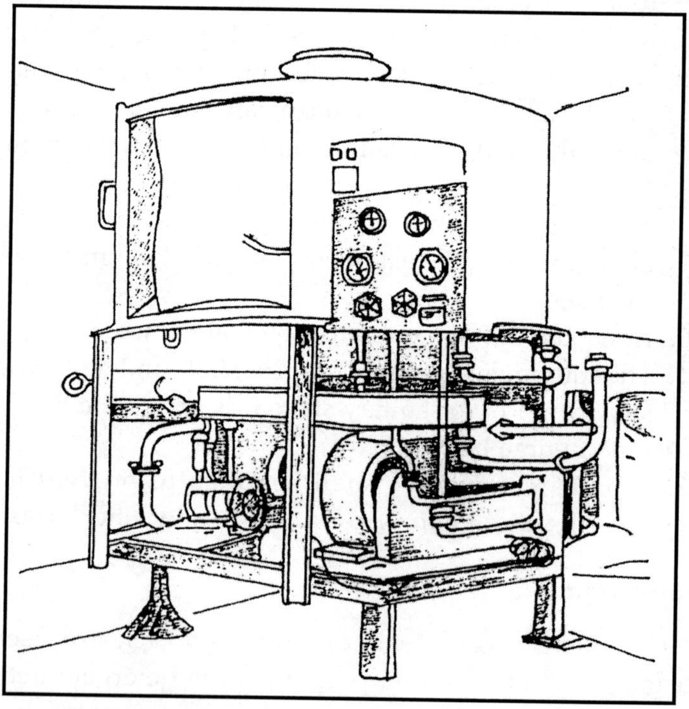

Fig. 11.1: Rotury can water

Choice of Detergents

Before choosing the detergents the following points are to be looked into:
i) Economy, ii) Availability, iii) Solubility, iv) Emulsification property, v) Saponification value, vi) Non-corrosiveness, vii) Sequestering and chelating ability, viii) Wetting and penetration, ix) Defloculation, x) Free rinsing.

Properties of a Good Detergent

1. Water softening
2. Wetting and penetrating
3. Scale prevention
4. Suspension
5. Peptisation
6. Heat stability
7. Germicidal activity
8. Sequestering and chelating power
9. Saponifying and emulsifying power
10. Non-corrosive/non-irritation on skin.

Application

Since single detergent may not possess all the essential properties, it is always recommended to use mixture of more detergents–compounded for different operation.

Dairy Detergents

According to their chemical nature, and to the various purposes for which they are intended, the following types of detergents are generally used in dairy plant.

1. *Alkali:* Most commonly used alkalis are sodium hydroxide, sodium carbonate, sodium phosphate, sodium silicate, trisodium phosphate, sodium-hexametaphosphate. All these constituents possess an alkaline reaction-necessary

for dissolving casein and for emulsification and solution of fat besides maintaining germicidal properties. Both strong and weak alkalis are used. A mixture of 60 parts soda ash, 25 parts trisodium phosphate, and 15 parts sodium metasilicate is used as general cleaning agent in dairy plant. The inclusion of 10 to 20 parts per million of calgon, in the mixture has proved effective in leaving a brilliant finish for washing of bottles. Trisodium phosphate has very good emulsifying powers, whereas sodium metasilicate possess good buffering properties.

2. *Mild acids:* Mild acids like phosphoric, citric, gluconic, tartaric, etc. are used to remove the milk stone, whereas strong acids like nitric acid is applied on stainless steel surface. The acid wetting agents are sanitising agent consisting of a mixture of phosphoric acid plus anionic and non-ionic wetting agents. It is usually employed in dilutions of 200 ppm of active ingredient. It is sufficiently germicidal and has been found to kill *E. coli or S. aureus* within 30 secs. at 70 to 75° F (21 to 24°C). Phosphoric acid diluted to 140 g/45.3 kg is generally used for dairy utensils.

3. *Water softeners:* Different water softeners are used to combat the situations arise out of using hard water in dairy plant (scale/milk stone/water stone, etc). Such softners are tetra phosphate, hexametaphosphate, tri-poly phosphate, etc. which are used in combination with alkalis or acids.

4. *Synthetic detergents:* These are good emulsifier and surface active agents and have good penetrating and wetting ability because of their capacity in lowering surface tension of water. Teepol, common soap, acinol-T are used as surface-active agent either singly or in combination with acids or alkalis.

5. *Sanitisers:* Sanitisation refers to the process of treating a utensil or piece of dairy equipment with physical or chemical agents in order to destroy most but not necessarily all, of the microorganisms present.

   Sanitisation of dairy equipment is accomplished most often by application of chemicals known as germicides (germ killers), bactericides (bacteria killer), viricides (virus killer), fungicides (fungus killer), etc.

Sanitisers are substances which are capable of destroying all pathogenic and most of the non-pathogenic organisms.

The desirable properties of a sanitiser are:

1. Non-toxic
2. Non-corrosive to hand and equipment
3. Cheaper
4. Easily and quickly applied
5. Effective in action.

Effective sanitisation can be achieved by:

i) Steam/heat, ii) Hot water, iii) Radiation, iv) Chemicals.

Steam/Heat

Heat in the form of steam under pressure is employed in system where complete sterility is required as in aseptic canning of filling process. Heat for sterilisation of equipment may be applied in the form of flowing steam or hot air. Sanitisation with steam is applied for sanitisation, it is desirable that it is to be forced in to a closed system and sufficient steam must be supplied so that the condensate at the outlet end of the system runs no lower that 200°F for a 5 minute period.

Hot Water

Immersion of small components like knives, small parts utensils, and small containers into water heated to 75°C to 80°C or higher is another thermal method of sterilisation.

For hot water to be effective as a sanitiser, it must be heated high enough so that its temperature at the outlet end of the system is maintained at a minimum of 170°F (77°C) for a continuous flow period of at least 5 minutes.

The microbicidal action is thought to be due to denaturation of protein molecules in the cell. Hot water is an effective non-selective sanitising method for food contact surfaces. Generally 15 minutes exposure time at 85°C or 20 minutes at 80°C is required.

Radiation

Radiation in the form of ultraviolet, high energy cathode, or gamma rays rapidly destroys microorganisms. UV ray in the wave length of 2537 °A has been used extensively in the form of sterilising lamps to destroy undiserable organism in food and dairy plants. Gamma and cathode rays have been applied successfully for sanitisation of materials such as surgical supplies.

Chemicals/Chemical Agent

A variety of chemical compounds is available for preparation of sanitising solutions. These vary greatly in chemical composition and activity under different conditions. Chlorine compounds like liquid chlorine, hypochlorites, inorganic and organic chloramines function as sanitisers. These compounds kill the microbes by inhibiting carbohydrate metabolism.

Sodium hypochlorite is the most widely used chemical sanitiser in the dairy industry, chiefly because it is readily soluble, inexpensive, active against a wide variety of microorganisms, and unaffected by hard water salts (200 ppm).

The chlorine-releasing products have rapid activity in destruction of bacteriophage.

Two types of iodine sanitisers are commonly used in dairy plant.

1. Iodophor-consists of a complex of non-ionic wetting agent.
2. Iodine, loosely bound.

The compounds are capable of rapid destruction of a wide variety of non-spore forming bacteria and at the same time not affected by hard water salts. Thus they are useful for sanitising most types of dairy equipment (25 ppm, pH 6.0 or lower).

Quarternary ammonium compounds have been found to exert pronounced bacteriostatic action which is of distinct value for certain dairy sanitising operations. They are also non-irritating to skin surfaces, more stable in presence of organic matter than some other germicides.

Factors influencing action of a sanitiser:

i. Selectivity of sanitiser
ii. Concentration
iii. Time allowed for action
iv. Individual characteristics of the organisms
v. Temperature
vi. Effect of pH
vii. Nature of organic matter and milkstone
viii. Surface tension.

Procedure of Cleaning and Sanitising

Steps of cleaning and sanitising

1. Draining: To remove any residual loose milk and other matter.
2. Pre-rinsing: To remove milk residue sticking to the surface including other adhering materials.
3. Detergent washing: Washing with hot alkaline detergent solution to remove remaining milk solids if any.
4. Hot water rinsing: To clean the surface perfectly and to remove residual detergents.

5. Applying sanitiser: To destroy all pathogens and most of the non-pathogens.
6. Draining and drying: To prevent corrosion and to make the material surface unfavourable for bacterial growth.

Methods of Cleaning and Sanitising

1. Hand washing: In organised dairy normal cleaning and sanitising of hand washed equipments and glass bottles are done in the following way: Preparation of detergent solution →thorough rinsing of utensils in cold water → thorough brushing of equipments with detergent → cold water washing → draining and drying → sanitising by steam/hot water after rinsing with chlorine solution.
2. Mechanical washing: Mainly cans and bottles and trays are washed mechanically.
   Steps: For can washing: Draining for liquid milk residue → pre-rinsing with cold/luke warm water → draining → application of detergent→ draining→ rinsing with hot water→ or by steam → fresh water washing with hot water/steam → hot air drying.
   Stages of mechanical bottle washing
   Pre-rinsing → detergent wash (sodas with chelating and wetting agent) → rinsing with warm water → cold water rinsing (with chlorine solution, 200 ppm) → draining.

Cleaning in Place (CIP)

The system of cleaning and sanitising dairy plant without daily dismantling of equipment is popularly known as CIP system. This system ensures that all units of equipments receive uniform treatment day after day. It saves clean up cost by 25 per cent compared to manual cleaning and prevents damage that may be caused by dairy dismantling. It also reduces possibility of human error in contamination and assembly.

Effective CIP cleaning depends on:

1. Proper engineering and installation
2. Temperature and concentration of cleaning solution
3. Adequate velocity of cleaning solution
4. Use of proper chemical
5. Sufficient time.

Steps of In-Place-Cleaning of HTST Pasteuriser

Cold water rinsing → circulation of hot acid detergent (30 minutes) (commercial phosphoric acid 70-72°C → flushing with hot water (68°C for 15 min) → circulation of hot alkaline detergent (caustic soda; 82 to 83°C for 30 min) → Final rinsing with water (88 to 90°C)

Time schedule

Cold water	15 minutes
Acid detergent	30 minutes
Hot water	15 minutes
Caustic detergent	30 minutes
Scalding	15 minutes

Cleaning and Sanitisation

Standards set for cleaning in place (CIP) program are:

i. Strength of acid	1.0 per cent
ii. Strength of caustic	0.7 to 0.8 per cent
iii. Temperature of acid solution	60 to 65°C
iv. Temperature of caustic solution	60 to 65°C
v. Temperature of hot water	80 to 90°C

CIP Cleaning of Equipments and Store Tank Sequence

i. Plain water	10 minutes
ii. Caustic solution	20 to 25 minutes
iii. Plain water	10 minutes

iv. Acid	20 to 25 minutes
v. Plain water	10 minutes
vi. Hot water	25 minutes

Cleaning solution for butyrometer and other glass ware: Butyrometers and other glass ware which become difficult to clean can be cleaned by immersion them in the following cleaning solution for 12 hrs or more.

Cleaning Solution

Potassium dichromate	60 gm
Water	1000 ml
Commercial sulphuric acid	60 ml.

Cleaning of Glass Slides

1. Boil the slides for 10 to 15 min in a solution of soft soap containing a little of potassium permanganet → wash → dry with clean water → immersed in an equal quantity of 96 per cent alcohol and conc.
2. Nitric acid in a container for one hour → washed in running water for 15 min. → stored in 50 per cent alcohol.
3. Cleaninig of bacteriological apparatus (glass wares) boiled in soft soap solution for 30 min → washed → immersed in solution of dilute hydrochloric acid for 10 min. → thoroughly rinsed → dried.
4. *Cleaning of milk storage tank:* Pre rinse with tap water → draining → post rinse with hot water → sanitise with chlorine solution → draining.
5. *Cleaning of dairy floor:* Flush the floor with tap water → detergent washing → scrubbing the floors with floor brush → wash with luke warm water → draining the water → dry by using the floor squeezy.
   Detergent for greasy floors:
   Soda ash –70 per cent
   Trisodium phosphate- 10 per cent

Sodium metasilicate- 15 per cent

Soap flakes – 5 per cent

6. *Cleaning of HTST pasteuriser*: Disconnect various pipelines → put them in part water → open the past chamber → clean them with detergent solution → clean pipelines using brush and detergent solution → circulate hot water through the pipelines and the plates → assemble all parts → check for leakage → sterilise the plant by chlorinated water (200 ppm) or scalding water 20 to 30 min.

Determination of chlorine content of sterilising solution: Take 50 ml of solution to be tested in a flask → add and dissolve 1 to 2 g of crystal of potassium iodide → add 10 ml of 99.5 per cent acetic acid → free chlorine liberated produces yellow colour in the solution by liberating equivalent amount of iodine from potassium iodide → titrate with N/10 sodium thiosulphate until yellow colour disappears → add 2 ml of 2 per cent starch solution as an indicator → blue colour disappears.

The available chlorine = No of ml of N/10 sodium thiosulphate × 71 (in parts per million).

CHAPTER 12

Utilisation of Milk By-products

Dairy by-product is a product of commercial value produced during the manufacture of the main product. Economic disposal of by-products is an essential pre-requisite to profitable dairying because of their great and unique nutritional value. It should be noted that conversion of edible substance into non-food items is not ordinarily justifiable in India, where there is an overall shortage of milk supplies.

Table 12.1: By-products of Indian dairy industry

Main product	*By-product*
Cream	Skim milk
Butter	Butter milk
Ghee	Lassi, ghee residue
Chhana, cheese, casein	Whey

Table 12.2: Composition of by-products (%)

Name of by-product	*Water*	*Fat*	*Protein*	*Lactose*	*Ash*
Skim milk	90.6	0.1	3.6	5	0.7
Butter milk	91	0.4	3.4	4.5	0.7
Lassi	96.2	0.8	1.4	1.2	0.4
Chhana whey	93.6	0.5	0.4	5.1	0.4
Cheese whey	93.1	0.3	0.9	4.9	0.6

Skim Milk

On an average, 85 lbs of skim milk are available from 100 lbs of milk separated. The question of the utilisation of this skim milk arises whenever milk is separated to obtain cream for market purposes or butter making.

Skim milk as an animal feed: Farms engaged in marketing of cream or butter, utilise a portion of skim milk for raising the calves and the remaining portion is used for feeding pigs and poultry.

Skim milk for human food: About one-half of skim milk produced is utilised as a source of animal protein in the human diet.

Table 12.3: Other use of skim milk

Principle of utilisation	*Food products made*
Pasteurisation	Flavoured milks
Sterilisation	Sterilised flavoured milk
Fermentation	Cultured butter milk, acidophilus milk
Fermentation and concentration	Concentrated sour skim milk
Concentration	Plain condensed skim milk, sweetened condensed skim milk
Drying	Dried skim milk
Coagulation	Cottage cheese, baker's cheese, casein, sapsago cheese.

Flavoured milk: Flavours are added to the milk for profitable utilisation of skim milk by making the milk more palatable. It may be chocolate or fruit flavoured.

Flow diagram of flavoured milk preparation: Milk—standardisation—pre-heating—homogenisation—mixing flavour, colour, sugar, stabiliser—pasteurisation—cooling—bottling and storage.

Cultured butter milk and acidophilus milk: These are fermented milk products where selected microorganisms or starter culture are employed to develop the characteristic flavour and/or body and texture of the product. Generally, lactic starter is used for this purpose. These products are much more palatable than milk, easily assimilated, highly nutritive, and contain beneficial antibiotics.

Flow diagram of preparation: Skim milk—filtration—homogenisation— pasteurisation—cooling—inoculation of starter culture—incubation—coagulation—break-up of coagulum—cooling—packaging and storage.

Concentrated sour skim milk: By lactic fermentation about 2 per cent acidity is produced in the skim milk, which is then concentrated to about one-third to its weight. It is done for good keeping quality, smoother texture, and more stable product.

Flow diagram of preparation: Skim milk—pasteurisation—cooling—inoculation of starter culture—incubation—breaking of coagulum—concentration by evaporation.

Condensed skim milk (plain or sweetened): Skim milk is condensed to 2.4:1 to 4:1 and may be unsweetened or sweetened by sugar. It is used in ice cream factories and bakeries. Its keeping quality is similar to high quality pasteurised milk.

Flow diagram of preparation: Skim milk→filtration→standardisation→pre-heating→addition of sugar (only in case of sweetened condensed skim milk)→condensing→homogenisation→cooling and crystallisation→packaging.

Dried skim milk: Dried skim milk or skim milk powder is used for preparation of toned, double toned, recombined milk, tea, coffee, confectionery, ice cream manufacture, culture products, etc.

Flow diagram of preparation: Skim milk→filtration→pre-heating →separation/clarification→condensing→pre-heating→ drying→packaging.

Cottage cheese and baker's cheese: Cottage cheese is a soft, unripened cheese with mildly acidic flavour, consisting of flakes of curd having meaty consistency. Baker's cheese is similar to cottage cheese but softer and finer grained.

Flow diagram of preparation: Skim milk→addition of calcium chloride→addition of starter→addition of rennet→setting→ cutting→cooking→drainage of whey→washing and draining of curd→salting→creaming→packaging.
Cooking and washing is not done in case of baker's cheese

Sapsago cheese: Sapsago cheese is a very hard, cone shaped, greenish grating cheese, which derives its flavour and colour from the leaves of an aromatic clover mixed into it.

Casein: Casein is of two types, viz. industrial and edible casein. Industrial casein may be acid casein or rennet casein. Acid casein is used for manufacture of plastics (buckles, buttons, jewellery, etc). Rennet casein is used for preparation of adhesive, paint, paper coating, leather and textile finish, medicine, bristles and fibres, etc.

Flow diagram of casein preparation: Skim milk→precipitation→ draining and washing→pressing→milling and spreading→ drying→grinding and packaging.

Edible casein is prepared by maintaining strict hygienic condition at each step and it is used for preparation of coffee whitener, ice cream whipping powder, and protein hydrolysate, etc.

Butter Milk

The composition and food value of butter milk is essentially the same as those of skim milk, but the fat content is somewhat higher. At present large quantities of butter milk is practically wasted or at best are not used economically. A common practice concerning the smaller creameries is to sell the butter milk to someone who uses it for feeding swine. Fed in reasonable quantities butter milk is an excellent supplement of cereals and corns.

Commercial products from butter milk: Butter milk is used as condensed butter milk, dried butter milk and soft cheese. Condensed butter milk is a semi-solid where butter milk has been ripened and condensed to 3:1 (approximately). Sour or sweet cream may be used for this purpose. Condensed butter milk is also used for feeding pigs and poultry. (Dried butter milk or butter milk powder has been discussed under dehydrated milk product in this book). Due to its economical advantages the practice of drying butter milk is rapidly coming into use. The dry product proofs to be especially valuable as a constituent in the ration of young chicks. It is also used advantageously in calf raising as a substitute of skim milk. A small amount of butter milk is used in the manufacture of casein. Condensed and dried butter milk has some especially desirable qualities and are used exclusively in connection with the preparation of human food including ice cream.

Flow diagram of soft cheese preparation: Butter milk→addition of skim milk→pasteurisation→cooling→inoculation of lactic starter→cutting→cooking→pressing→packaging.

Whey

Whey is obtained as a by-product of cheese industries. The solids in true solution in milk largely pass into whey in the manufacture of cheese or casein. It is estimated that of the total solids in milk, 42 to 44 per cent are to be found in whey. These solids include over 90 per cent of the milk sugar and lactalbumin, a portion of the mineral matter and fat, as well as a very high per centage of the water-soluble vitamins of the original milk.

Commercial use of whey

Whey beverage: Whey is used in the production of fermented beverages both alcoholic and non-alcoholic. Whevit, a soft drink is such type of beverage.

Flow diagram: Whey→steaming→cooling→filtration→addition of sugar syrup→citric acid→sugar mixture, colour, flavour→ inoculation of culture→incubation→pasteurisation→cooling→ storage.

Yeast whey: Production of yeast from whey as a food supplement rich in protein and vitamins is another utilisation of whey.

Condensed whey (plain or sweetened): Pasteurising the separated whey and draining the hot liquid into the vacuum pan produce this. Sugar is added to the equal weight of the solids in the whey in case of sweetened condensed whey. Plain condensed whey may be used in human and animal food products, in confectionery, baked food, and cheese food and makes excellent chicken feed. Sweetened condensed whey may be used in fruit whips, certain candies, and frozen dessert preparation.

Whey protein concentrate: This includes soluble and coagulated whey protein. A coagulated whey protein is recovered by heat denaturation and acid precipitation, followed by filtration. Partially neutralising separated whey and then condensing, cooling, stirring, and centrifuging the same soluble whey protein is recovered. Both this whey protein may be used in baking, baby foods, ice cream mixes, whipping agents and emulsifying agents, etc.

Lactose: The most important source of lactose is cheese whey, although hydrochloric acid-casein-whey is still considered to be the best source for its production. In India chhana whey is the greatest source of lactose. Lactose has multifaceted uses such as preparation of humanised milk, infant food, caramels, silvering mirrors, bakery products, preparation of pills, and tablets.

Flow diagram: Whey→evaporation→crystallisation→separation of impurities→lactose crude→refining→drying→milling→ packing and storage.

Mysost and primost cheese: These are Scandinavian products. Mysost is made from cow-milk whey and it is darker brown and have a coarser texture. Primost is obtained from whey containing additional fat and it is light tan in colour with a caramelised flavour and a smooth creamy body.

Ghee-residue

Ghee-residue is a nutritious food having denatured protein, fat, burnt lactose and minerals. It can be utilised in preparation of candy and chocolate; as a paste as sandwitch spread, etc. The excess fat from this is removed in following ways:

Ghee-residue→enclosing in muslin cloth→boiling in water bath→draining→cooling→squeesing.

CHAPTER 13

Role of Milk and Milk Products in Human Nutrition

For centuries, milk has been recognised as almost perfect food for the mankind. Previously all the foods were looked upon merely as source of proteins and energy, but nowadays attention is given to the fact that milk contains all the food constituents required in the human diet and essentially with proper properties. Milk becomes an important food for growing children, pregnant mother, adults, by supplying these body building protein, energy giving lactose and fat, health giving vitamins, and bone forming minerals in fairly good proportion in following ways:

1. *Milk as a source of water:* Human being cannot live without water. All tissues contain water in varying amount. Water holds various solids in suspension and serves a carrier of food and waste in the body. It also helps to regulate body temperature. Milk consists of a relatively high per centage (86-88%) of water. So milk facilitates all the above said function by providing water to the human body as a major constituent of milk.

2. *Milk as a source of protein:* Nowadays it has been well established that the human being can survive for a long period without fat or carbohydrate, but without certain

proteins they will die shortly. Dietary proteins are major source of body building components and they synthesise different proteineous material and structures of the body. Human being cannot synthesise certain amino acids in the body, which are called as essential amino acid, and they should be supplied through diet. Milk contains all these essential amino acids, viz. tryptophan, phenylalanine, lysine, leucine, threonine, valine, isoleucine, methionine, arginine, and histidine in fairly high amounts. For this reason milk proteins are considered to be superior to vegetable protein. For the vegetarian it is the only source of animal protein. In addition the milk protein, casein and albumin are associated with some important minerals like phosphorus and calcium as well as some vitamins, which seldom occur in protein from other sources. Milk is comparatively cheaper source of protein supplying animal foods like egg, meat and fish. The tissues damaged by normal wear and tear are repaired and maintained by milk proteins. Milk protein also imparts immunological properties due to its lactalbumin content.

3. *Milk as a source of carbohydrate:* The principal carbohydrate found in milk is milk sugar or lactose. Certain amount of glucose is also present in milk. Lactose produces glucose and galactose during digestion. These are generally utilised in the body for the production of energy and heat. Milk is superior to other naturally occurring carbohydrate food source because it only contains lactose. Lactose facilitates assimilation of calcium and produces mildly acidic medium in the intestine which checks the growth of some proteolytic bacteria. Galactose helps in the development of cerebrocides.

4. *Milk as a source of lipid:* Lipid being an ester of fatty acids and related compounds functions as a source of energy production in the body. Milk fat plays a significant role

in the physical properties of milk and milk product, flavour, and nutritive value of milk. It contains essential fatty acids like linoleic acid and arachidonic acid and fat soluble vitamins like A, D, E and K. For these reasons it is superior to other animal fat. In dairy products, milk fat provides a soft body, smooth texture, rich taste, and characteristic flavour, which help in better acceptability of these products to consumers. Recently health conscious people tend to avoid milk fat consumption due to presence of cholesterol, which is responsible for heart disease.

5. *Milk as a source of minerals:* Milk supplies various inorganic elements, which are essential in human diet. The major salts present in milk are calcium, sodium, potassium, magnesium, phosphorus, citrate, bicarbonate chloride, sulphate, etc. The trace elements are aluminium, cobalt, copper, iron, lead, molybdenum, barium, silicon, zinc, etc. All these minerals are required for many metabolic processes in the body. Milk is an excellent source of calcium and phosphorus, both of which together with vitamin D are essential for bone and teeth formation. Calcium and phosphorus remain in the milk in a highly satisfactory ratio (1:43:1) and in an adequate quantity (1.185 g calcium and 0.88 g phosphorus), required by growing children and pregnant women.

6. *Milk as a source of vitamins:* Vitamins are accessory food factors required in small amount for satisfactory growth, health and reproduction. They have to be supplied in the diet as because they are not synthesised in the human body. Milk contains fat soluble vitamins like A, D, E, and K and water soluble B-complex like B1, B2, B6, and B12. Vitamin A is required for general growth, health of the eyes, structure, and function of skin and mucus membrane. Thiamine (B1) helps in growth and function of nervous system, appetite, and prevents beriberi.

Functions of stomach, intestine, and nervous system are facilitated by niacin. Vitamin D aids in calcium absorption, which strengthen bones and prevents rickets. Milk is deficient in vitamin C. Vitamin K is essential for blood clotting.

7. *Milk as a source of energy:* All the milk solids except ash serve the body as sources of energy. The energy value of milk varies with its composition. Cow milk provides 75 calorie per 100 g and buffalo milk provides 100 calorie per 100 g of milk on an average. Milk fat contributes 9 calorie whereas protein and carbohydrate provide 4 calorie energy per g. Milk supplies these calories relatively at cheaper rate than other animal foods like meats, fish, and eggs.

Food Value of Milk Products

1. *Butter:* Highly concentrated fat food, contains very little amount of vitamin D and E, lacking vitamin B1 and C. It is an excellent source of energy and vitamin A (approximately 15000 IU/lb).
2. *Ghee:* Richest source of milk fat, very low in fat-soluble vitamin A and D.
3. *Ghee-residue:* Rich source of milk proteins and minerals.
4. *Cheese:* Food value depends on the proportion of water retained with milk solids. It is the cheapest source of animal protein adopted for human food; is an excellent source of calcium, phosphorus, and sulphur; fairly high in iron and fat soluble vitamin like A and D. Owing to its high protein and low sugar content cheese is highly recommended for diabetic patient.
5. *Chhana:* Important acid-coagulated indigenous milk product similar to soft cheese.

6. *Dahi:* Easily digestible and assimilable product; seems to exert a possible therapeutic value in the stomach and intestinal disorders due to desirable bacteria that establish themselves in the lower intestine and compete with the undesirable putrefactive, toxin producing bacteria.
7. *Lassi:* Contains appreciable amounts of milk proteins and phospholipids and is an excellent beverage for quenching thirst by reason of its lactic acid.
8. *Concentrated milk:* The food value depends on the extent to which they are concentrated (e.g. kheer = 2 folds, khoa = 4 folds, rabri = 5 folds concentration).
9. *Ice-cream:* Food value varies with its composition. However, all the constituents of milk are present in a concentrated form (60-68% water). In addition, it contains sugar, chocolate, bakery products, gelatin, eggs, fruits, nuts, etc.
10. *Skim milk:* Since it contains a fraction of milk fat, the vitamin A, D, and E content is low. The water-soluble vitamins are found in skim milk. It is an excellent source of animal protein of high quality.
11. *Butter milk:* Comparable with the skim milk excepting that buttermilk contains slightly more milk fat. Due to the decomposition of lactose by bacteria, butter milk frequently contains 0.2 to 1 per cent of lactic acid, which is not harmful and may be beneficial.
12. *Toned milk:* (Single toned milk) refers to milk obtained by the addition of water and skim milk powder to whole milk or removal of excess fat in the form of cream from whole milk by means of separation of butter oil with skim milk powder. The SNF level is same with the whole milk but here the fat content is reduced.

Constituents	*Standardised*	*Toned*	*Double toned*	*Skim milk*
Milk fat	4.5	3.0	1.5	0.5
SNF	8.5	8.5	9.0	8.7

CHAPTER 14

Milk and Public Health (Milk Borne Diseases)

Milk can be a potential carrier of disease producing organisms. Unless proper precautions are taken, out-breaks of milk borne diseases can occur. Diseases which are known to be transmissible through milk are discussed below along with their manner of entry into the milk.

1. *Infection of milk directly from the cow:* The causative organisms of bovine diseases enter the milk through the mammary gland or through faecal contamination, and may cause a disease condition in persons who consume the milk, e.g. bovine TB, undulant fever, anthrax, cow pox, etc.
2. *Infection from man to cow and then to milk:* Some of the human diseases can become established in the cow's udders, e.g. septic sore throat, scarlet fever, diphtheria, etc.
3. *Direct contamination of milk by human beings:* By direct human contact, either carriers or patients, some diseases may be transmitted to the milk, e.g. septic sore throat, scarlet fever, typhoid fever, dysentery, gastroenteritis, etc.
4. *Indirect contamination of milk by human beings:* The pathogenic organisms of some diseases may enter the milk through contaminated bottles, utensils, water supply, insects, and dust, e.g. typhoid, paratyphoid, diarrhea, etc.

CONTROL AND PREVENTION MEASURES

1. Milk should be drawn only from healthy animals. Animal should be periodically tested for the contagious and infectious diseases specially TB and brucellosis and all diseased animals should be removed from herd.
2. Adequate measures should be taken to restrict the movement of human beings in the animal shed so that the transmission of human diseases to the animal can be prevented.
3. Care should be taken to avoid direct contact of carriers or patients during handling and processing of milk. Health of employees should be safe guarded by regular medical examination.
4. All the dairy equipments, utensils, and water supply should be sterilised properly and free from any pathogenic organisms and dust, etc.
5. Care should be taken to control the flies in the plant.
6. Milk should be properly pasteurised and secondary contamination must be avoided after heating, during cooling, and bottling.
7. Pasteurised milk should be placed in sealed container and stored at a sufficiently low temperature to prevent the multiplication of microorganism that may have survived pasteurisation or that may have contaminated the milk subsequent to the heat treatment.
8. A high standard of sanitation should be maintained in the farm, processing plant, during transport, and in home upto consumption. As essential control measure that should be undertaken now is the education of milk producers and public for the hygienic handling of milk.

CHAPTER 15

Synthetic Milk

Synthetic milk actually is a misnomer. It is basically a conconction, adulterating natural milk with pulverized soap, vegetable oils, caustic soda and urea. Such is the accurate blending that the adulterated milk appears smells and tastes no different from natural milk. In majority of cases detected, the preferance was for refined palmolein oil. In very few cases, (one sample from Rajasthan) have been found to contain even kerosene oil. According to earlier report of ICMR, adulterants like caustic was found to have a carcinogenic effect on the human body. (The Times of India, Ahmedabad 07.06.96 Devinder Sharma).

Environmental pollution of air, water or soil is gradually reaching to its peak due to increase in industrialization, vehicular influx in big cities, fertilizers and pesticides utilization in crops, but the main threat in this decade has been from man-made adulteration in human food items, particularly in milk. The menance of substitution of synthetic components to the natural constituents of milk has crossed all the limits.

Components like neutralizer, urea, formaline stabilisers, detergents, sub-standard mineral oil etc have been used by uncruplous trades regularly to make synthetic milk.

Synthetic milk is a gross misuse of scientific knowledge of the process of saponification. Synthetic milk is synthesised by mixing liquid or pulverised soap with vegetable oil and water. Adjustment of non-fat solids in synthetic milk is carried out by adding common salt, sugar urea, etc.

A less harmful variety of synthetic milk is produced by simply replacing the butter fat by vegetable oil. The product is also homogenised by mixing vegetable oil thoroughly so that the adulteration is not detected on external appearance. (The Hindustan Times, New Delhi, 29.1.96 Anil Maheswari).

According to estimates, synthetic milk costs some what around Rs. 2 per litre, whereas, real milk costs more than Rs. 12 per litre. Because of the large difference in price, many people are tempted to buy it. Further, the powder was being used in some states especially in and around Calcutta, for making sweets (as reported by Mr Sharma. The Asian Age, Mumbai/Londo 01.06.96). It has further been reported that although the manufacture and sale of synthetic milk was rampant in north India, the 'poison' was gradually spreading to the other parts of the country.

With the help of lactometer, the uncrupulous traders adjust the specific gravity of the adulterated milk to that of pure buffalo milk. And that is perhaps the reason why even the milk plants find it difficult to detect synthetic milk at the time of procurement. It is, however, another matter that many of the private milk plants are themselves involved in adulteration.

It has also been reported (Health Tribune, Wednesday Feb 11, 1998, By V K Garg) that in several parts of the country milk vendors are "selling synthetic milk" which is adulterated with harmful chemicals.

Synthetic milk is an emulsion which is made by mixing cheap quality cooking oil and detergents in water. Other ingredients of synthetic milk are urea, caustic soda and powdered sugar and skimmed milk powder. Detergents are added to make cooking oil soluble in water. Caustic soda is added to prevent it from turning sour. Synthetic milk resembles natural milk in all respect but has no nutritional value and is poisonous.

Synthetic milk have long-term ill effects on the health of an individual. Urea and caustic soda are harmful to the heart, the liver and the kidneys. Caustic soda acts as slow poison for those suffering from hypertension and heart ailments. Caustic soda also interferes in the digestion of amino acids present in proteins and make them indigestible. Urea is an additional burden for the kidneys as they have to do more work to remove urea from the body. Synthetic milk is harmful for all, but is more dangerous for pregnant women and neonatals.

Density (0.1) and fat contents (4.5%) of natural and synthetic milk are almost same. But fat contents of synthetic milk can be varied easily by increasing or decreasing the percentage of the cooking oil in it. Natural milk adulterated with synthetic milk differ in taste and turns yellowish after some time.

No single test is available presently to detect adulteration of natural milk with synthetic milk at the consumer level. Presence of urea can be detected by using paradimethyl amino benzaldehyde (DMAB), test. Urea gives yellow colour with DMAB. But as urea is present in natural milk also (although to a smaller extent than synthetic milk), so this test may be confusing for non-technical consumers. The cheap cooking oil of synthetic milk (but yet no simple test) cannot easily be distinguished from fats of natural milk. To detect this cooking oil in the milk sophisti-equipments are required which is again not a practical solution to the problem.

Lactometers also can not detect chemical adulteration of milk as they do not detect its chemical constituents. Detergent is the only ingredient which is exclusive to the synthetic milk but yet no simple test is available to detect. The National Dairy Research Institute, has reportedly found simple methods to detect the adulteration of milk. In this regard a simple kit is yet to be commercialised.

Table 15.1: Main characteristics

Properties	*Natural milk*	*Synthetic milk*
Colour	White	White
Taste	Palatable	Bilter
Effect of heating	Remains white on boiling	On boiling the colour becomes yellowish
Effect of storage	No change in colour	Turns yellowish after sometime
Texture	No soapy feeling, if rubbed between fingers	Gives a soapy feeling if rubbed between fingers.
Test of urea	Weakly positive (light yellow)	Highly positive (intense yellow)
pH	Neutral (6.8)	Alkaline (10-11)

Some physicochemical parameters like pH, neutraliser thickner (CMC), acidity (before boiling and after boiling) protein percentage, fat/SNF contents, sugar contents have been studied. (17-brands of milk - 25 types in available in the Delhi market). pH was measured by portable digital pH meter. Other determinations have been done by common chemical methods. The results are tabulated in Table 15.1.

It has been found that except DMS milk only other four branded milk (e.g. J.B; Rama, Mother's and Jagriti) have good or normal taste. The other branded milk have poor or bad taste. The other parameters like pH, neutraliser, thickner (CMC), acidity (before and after boiling) protein percentage, fat/SNF contents, sugar contents have been determined. It has been found that the pH value varies from 6.76 to 7.11; acidity varies from 0.05 to 0.12 (before boiling) and from 0.03 to 0.11 (after boiling). Protein percentage varies from 2.6 to 3.8; fat content varies from as low as 0.15 to 3.05 max. SNF varies from 6.6 to 9.0. Neutraliser in 12 samples and sugar in 5 samples have been found to be positive. In few cases, the presence of detergent have been detected.

Table 15.2: Typical characteristics of synthetic milk prepared in the laboratory

Particulars	*BR*	*Protein*	*Acidity before boiling*	*Acidity after boiling*	*Neutra-liser*	*Taste*
SMP 8 gm + Dextrose - 1 hm + Sunflower 6 ml + Liquid soap 6 ml + Water 75 ml	61.5	2.44	0.04	0.03	Positive	Very abnormal
Pure milk	41.4	3.9	0.14	0.12	Negative	Normal
10% Synthetic milk + 90% Normal milk	42.6	3.6	0.12	0.10	Negative	Poor Taste
25% Synthetic milk + 75% Normal milk	43.2	3.2	0.10	0.08	Negative	Poor Taste

(Study of physiochemical parameters of natural, adulterated milk, AK Sarkar, KC Sharma, N Singh, BB Garg and BN Bhandary. P-54-57. XXX Dairy Industry Conference December 2000)

Table 15.3: Study of typical characteristic of synthetic milk production with vegetable fat, dextrose and urea, etc.

S. No.	*Particulars*	*Taste*	*BR*	*Protein*	*Acidity before boiling*	*Acidity after boiling*	*Neutra-liser*	*Urea*	*Sugar*
1.	Pure milk	Normal	41.4	3.8	0.11	0.10	Negative	Negative	Negative
2.	Milk 100 ml + Liquid soap 10 ml + Veg fat 1%	Poor	42.2	3.6	0.11	0.10	Negative	Negative	Negative
3.	100 gm milk + 1 gm urea + 1 gm dextrose + 0.5% veg fat + 0.5% liquid soap	Poor 2 Creamy	42.0	3.4	0.10	0.09	Positive	Positive	Positive

Contd...

Contd...

4. 100 gm milk + 2% dextrose + 0.5% teepole + 2.0% veg fat	Very bad	43.2	3.8	-	-	Negative	-	Slight yellow colour, serum content 2.5 times higher than normal.

Detection of adulteration and rapid tests the menance of synthetic milk can be checked.

Milk is generally recognized as an important source of nutrition for children as well as for adults. Hence adulteration of milk is a highly sensitive issue. The increasing incidence of sub. Standard milk is a matter of concern to dairies.

Recent reports in the media indicate that a new type of adulterated milk called synthetic milk has become a cause of serious concern for the Indian Dairy Industry. Synthetic milk shows all the properties of natural milk. Hence it is very difficult to detect synthetic milk. Moreover it was found to be harmful to the infants and children.

Uncrupulous agents have prepared formulations based on urea, detergents, etc. which have the appearance of natural milk. Not able to make good profit due to increasing cost of milk, some greedy traders have resorted to these harmful formulations to make quick money. To avoid adulteration of milk with the product of these formulations; it is necessary to develop a rapid and sensitive method to detect the presence of synthetic milk.

There are many methods known for detection of adulteration in milk. A simple, rapid and sensitive method has been developed to detect synthetic milk.

Earlier methods for detection of urea in milk take longer time (more than one hour). A formulation of synthetic milk was prepared in the laboratory incorporating urea, detergent and oil in water which can be used to adulterate milk, and the adulterated product could still pass the conventional tests of fat, SNF, MBR, etc.

The method for detection of synthetic milk is as follows:

Detection of Urea in Milk

Take 5 ml of milk in a test tube → add 0.2 ml of urease (20 mg/ml) → Shake well at room temp → then add 0.1 ml of Bromothymol blue solution (0.5%) → Appearance of violet colour indicates the presence of detergent in milk. Normal standard milk with this test shows faint violet colour.

Detection of Synthetic Milk

Take 5 ml of milk in a test tube → add 0.2 ml of urease (20 mg/ml) → Shake well at room temp → add 0.1 ml of Bromothymol Blue solution (0.5%) → Appearance of blue colour after 10-15 minutes. indicates the presence of urea in milk. Normal standard milk with this test gives faint blue colour indicating the urea present in the milk (0.02%).

Detection of Detergent in Milk

Take 5 ml of milk in a test tube → add 0.1 ml of Bromocresol Purple solution (0.5%). Appearance of violet colour indicates the presence of detergent in milk. Normal standard milk with this list shows faint violet colour.

Detection of Synthetic Milk

Take 5 ml of milk in a test tube→add 0.2 ml of urease (20 mg/ml)→shake well at room temperature→add 0.1 ml of Bromothymol blue solution (0.5 %)→appearance of dark blue colour indicates the presence of synthetic milk.

The detection of urea and synthetic milk are same, only difference in appearance of dark-blue colour in the case of synthetic milk.

To detect artificial milk made out of urea and detergent:

Scientists at Govind Ballav Pant Agriculture and Technology have devised an easy way to detect artificial milk made out of urea and detergent.

A simple paper strip test: when dipped in milk, the chemically treated paper strip turns yellow in proportion to the quantity of urea in the milk. The paper remains unchanged if the milk is pure.

New Method to Test Synthetic Milk

A scientist at the Haryana Agriculture University has developed a quick and easy test to detect the presence of synthetic milk (urea, etc):

"Put a drop of test solution on a filter paper → Pour equal measurement of the sample (milk) on it → A yellow ring around the milk drop will appear in few seconds if the milk is synthetic.

Urea was the basic ingredient of synthetic milk, which also contained detergents, caustic soda, vegetable oil, sodium sulphate and even foramalin and white paint. According to the researchers, the presence of urea and other ingredients in milk may cause damage to vital body organs, such as livers, kidneys and heart. Such milk was abundantly produced, in Uttar Pradesh and Haryana to meet short supply of pure milk during summer months.

SOME EMERGING AREAS OF INTEREST

1. Pesticide Residues in Milk and Milk Products

Pesticide means any substance intended for preventing, destroying, attracting, repelling, or controlling any pest, including unwanted species of plants or animals during the production, storage, transport, distribution and processing of food, agricultural commodities or animal feeds.

Pesticide residue means any specified substances in food, agricultural commodities or animal feed resulting from the use of a pesticide.

The term includes any derivatives of a pesticide such as conversion products, metapolites, reaction products, and impurities that are of toxic nature (IDF-9101, 1990).

Pesticide Residues in Milk

1. Organochloro pesticides (OCP) viz, DDT (dichrolo-diphenyl trichlorodiphenyl trichloroethane); BHC (benzene hexachloride); chlordane; heptachlor; aldrin; dieldrin, endrin, endosulphan, - are fat soluble, persistent and not readily excreted by the animals except in to milk.
2. Other pesticides such as organophosphates and carbamates have decisive advantages over the organochloro pesticides. These compounds are not fat soluble; much less persistent, degradable and excreted in urine, faeces and breath of animals and hence the build-up of significant residues in milk is prevented.

Persistent organochloro pesticides being sparingly soluble in aquous media are not readily available for microbial degradation or detoxification. However, being fat soluble, these remain stored in the animal body for long periods in an unchanged form. These pesticides deposits are metabolised at the time of lactation or stress and transported in to milk.

Sources of Pesticide Residues in Milk

The sources of pesticide residues in milk chain are represented below:

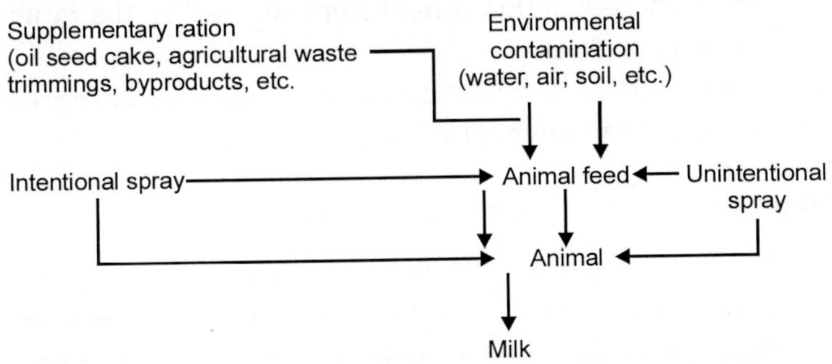

Table 15.4: Pesticide residue levels in milk and milk products in India

Product	*DDT (ppm fat basis)*	*BHC (ppm fat basis)*
Milk	0.01-55	0.02-61
Human milk	1.4-102.2	1.25-27.52
Butter	1.6-11.36	0.55-1.51
Ghee	0.25-7.24	0.30-23.80
Infant formula	0.17-4.30	0.12-5.70
Skim milk powder	0.80-1.24	0.43-0.78

Toxological Aspects

1. Cellular/genetic damage to plants, animals and human being.
2. Destruction of friendly insects - honeybees
3. Damage of peripheral nervous system, blindness
4. Tumorigenic effects
5. Liver functioning affected
6. Infant mortality
7. Acute poisoning

What can be done? preventive measures

1. Using more biodegradable pesticides/Biopesticides.
2. Encouragement of organic farming.
3. Rotation of crops, clean cultivation and water management.
4. adoption of genetically modified insect post resistant seed for commercial cultivation.
5. Safely and education must be encouraged in the rational use of pesticides.
6. Integrated Pest Management (IPM) and Integrated Nutrient Management (INM).

Other Potent Contaminants in Milk

1. *Residues of Antimicrobial Agents in Milk:*
   Residues of antimicrobial agents may be of toxological significance for the consumer and may influence on the

technological properties of milk for manufacture of fermented products.

2. *Contaminants in Milk - Dioxins and Mycotoxins:*

"Pest problems will become more serious with future intensification of agriculture. Therefore, farmers should move towards judicious use of less persistent and biodegradable synthetic pesticides and biopesticides, adoption of Integrated Pest Management, Integrated Nutrient Management and Organic farming, so that food needs could be met without encountering the pest problems and related contamination of food."

2. Layout of Milk Processing Plant and Its Management

1. Regional Considerations:
   a. Environment b. Attitudes
   c. People's demands d. Climate
2. Climatic Considerations:
   a. Normal maximum and minimum temperature in the hot season and cold season
   b. Humidity
   c. Other special conditions, e.g. severe seasonal wind/ severe rain fall.
3. Seasonal Considerations:
   a. Lean season
   b. Flush season
   c. Overall requirements
   d. Summer/Winter.
4. Milk Production-Trend Considerations:
   a. Extent of milk-shed
   b. Location of milk-producing areas and distance from the plant
   c. Total number of producers
   d. Average max and min daily production per day/per annum

e. Details of endemic diseases
f. Bacteriological and chemical quality of milk at point of production.
g. Cooling arrangement on farms.

5. Adoptable Practices Considerations:
   a. A stable source of good quality milk in required quantity at reasonable price throughout the year.
   b. Current testing system
   c. Practices at farm and factory level
   d. New methods of testing.
6. Milk Transport Considerations:
   a. Types and owenership of transport container used
   b. Transport facilities - road and rail
   c. State of roads throughout
   d. Max weight or size of truck/lorry permissible.
7. More Milk Production Target Considerations:
   a. To cover alowing loan for purchase of high yeilding animal
   b. Introducing Artificial Insemination and scientific breeding programme
   c. Making available feeds and fodder in subsidised rates
   d. Arranging veterinary aids to animals
   e. Extending financial help for animal housing
   f. Construction of silo
   g. Arranging sound water-supply system
   h. Arrangement for preserving milk
   i. Establishing educational facilities for scientific handling of animal, milk and cost minimizing concept.
8. Milk Utilization Considerations:
   a. Present forms in which milk is sold-liquid or as products
   b. Market potential
   c. Extent and characteristics of market area for liquid milk and for products.

9. Laboratory Accreditation Considerations:
   a. Testing methods used
   b. Chemical and microbiological analysis
   c. Quick testing system
   d. Purity testing methods.
10. Packaging and Distribution Considerations:
    a. Best form and size of container for liquid milk and for milk products
    b. Method of liquid sale
    c. Local storage required or not.
11. Services Facilities Considerations:
    a. Availability of fuels, relative cost and limitations on quantity.
    b. Availability of electricity supply, AC or DC, voltage and capacity
    c. Water supply quantity and quality
    d. Effluent disposal, details of public facilities and legislation.
12. Availability of Plant and Material Considerations:
    a. Availability of building materials in the local markets
    b. Availability of dairy equipments within the country
    c. Availability of other service equipments, e.g. boilers, refrigeration equipment, electric motors, pumps, etc.
    d. Availability of road tankers/vehicles.
13. Plant Organization Considerations:
    a. Availability of technical, clerical and unskilled staff
    b. Availability of ITI (Industrial Training Institute)
    c. Housing facilities for the staff.
14. Milk Quality Management Considerations:
    a. Microbes management
    b. Residues management
    c. Farm management
    d. Tropical milk management
    e. Animal health management

f. Animal feed management
g. Water management
h. Laboratory management
i. Analyst management
j. Man management.

3. Organic Milk Food Products

Milk food products obtained through biodynamic farming/ organic farming was advocated in Germany 1924 by Rudolf Steiner.

In a simple manner it emphasizes, milk production through organic farming. The organic milk production is a vital output of "Organic Farming"- a movement ushered in jointly by idealistic farmers and conscientious consumers.

Organic Dairy Production, a component of Organic Farming is basically characterized by following farm and live stock management practices:

1. Feeding of milch animals on clover and grass
2. Production of surplus Nitrogen by Nitrogen - Fixing Bacteria on the roots of clover
3. Disease free animals
4. Pesticide free feed and foodder
5. Produced by natural fertility of soil
6. Restricted or no use of feed concentrate
7. Pollution control/avoiding pollution
8. Preventive treatment of animal diseases and alternative therapy, e.g. homeopathy.

Special care should be taken for production of such organic foods, e.g. to exclude artificial or chemical ingredients including flavor, colour, sweetners, stabilizer, emulsifiers, etc.

Developing Organic Standards

Standards for Organic Food Production Methods have been formulated as regulations at national (regional) level, e.g.

European Union; International, e.g. codex; Private, e.g. national skal. Standards, Soil Association, Escort.

The formation of National Organic Standards and Certification Scheme based on European Union Standards are underway in India.

Some of the important guidelines are:

i. provide protection to consumers against deception and fraud

ii. extend protection to organic producers against mispresentation

iii. ensuring that organic food production, preparation, storage, transport marketing are liable to inspection and comply with the guidlines.

iv. harmonize provisions for the production certification, identification and labelling of organically produced food products.

Processing of Organic Milk

Some of the leading dairy organisations engaged in processing of organic milk world wide both in private and co-operative sectors.

1. MD Foods (Denmark): Country's largest organic milk producer/processor. It not only caters to the domestic need but also exports a large range of organic dairy products to neigh-bouring European countries.
2. Horizon Organic Dairy (USA): This company possesses two organic dairy farms and a herd of large number of cows. It markets all types of refrigerated milk products.
3. Robinson Dairy Inc (USA): It processes 3 to 5 million gallons of organic milk and dairy products annually.
4. Bioland (Germany): A co-operative organisation producing 114 million litres of organic milk annually
5. Naturemaelkamba (Denmark): This company processes only organic milk products-produces 9.5 million kg annually.

Organic Milk Production, an emerging trend in milk production seems to have generated a lot of interest among producers, processors, consumers, researchers and policy makers (Ref. IDM 53.12.2001).

4. Bactofugation

Bactofugation is the process of removing 99% of the bacteria in milk by centrifugal force.

Various methods have been used to remove bacteria centrifugally. High-speed clarification (20,000 rpm) has been found to remove 85 to 93% of the bacterial cells, depending upon the type of bacteria present. The density of bacteria varies from 1.07 to 1.3 which will affect the percent removal of a given species.

The 'bactofuge' consists of a clarifier with two small holes drilled in the bowl shell to allow continuous discharge of skim-milk. The clarifier is operated at about 6,000 rpm and at temperature of 77°C is used to reduce viscosity and increase efficiency.

In operation, generally two "bactofuges" are used in a series. The skim milk of the second machine is returned to the balance tank. The dual operation is claimed to remove 99.9% of the bacterial cells.

This process may be considered supplementary to pasteurisation, for it is still necessary to destroy the bacteria not removed.

References

1. Acharya RM, Chatterjee AK, Lya KK, Prahlad SN. Dairy India 5th Edition (1997). International Book Distributing Co. (IBD Co.), Lucknow, Mumbai.
2. Anantakrishnan CP, Khan AO , Padamanabhan PN. The Techology of Milk Processing, 2nd edition (1993). Sri Lakshmi Publishers, Madras.
3. Clarence, Eckles Henry. Dairy Cattle and Milk Production, 5th Edition (1956), Revised by Anthony, Ernest. L. Biotech Books, Delhi, 2001.
4. De, Sukumar. Outlines of Dairy Technology, 14th Edition (1996) Oxford University Press, Delhi, Bombay, Calcutta, Madras.
5. Eckles, Clarence Henry, Willes Barnes, Macy harold. Milk and Milk Products, 4th Edition (1973) Tata McGraw Hill Publishing Company Limited. Bombay, New Delhi.
6. Handbook on Animal Husbandry (2002) published by Indian Council of Agril. Research, New Delhi.
7. Harvey WM, Clunie and Hill, Harry. Milk Products nd edition (1999) Biotech. Books, Delhi 110035.
8. Jenness R and Patton. Principles of Dairy Chemistry (1970) Wiley Eastern Private Limited, New Delhi.
9. Malakar, Biman (1990) Contributory Chapter in Handbook of Animal Husbandry Science Edited by A Chakraborty, Kalyani Publishers, New Delhi.
10. Rangappa KS, Acharya KT. Indian Dairy Products (1974), Asia Publishing House, Bombay.
11. Sommer HH. Market Milk and Related Product. 2nd Edition. (1946).
12. Srinibasan MR, Anantakrishnan CP. Milk Products of India; ICAR Animal Husbandry series no. 4, New Delhi.
13. Webb Byron H, Johnson, Arnold H, Alford, John A. Fundamentals of Diary Chemistry, 2nd Edition (1987), CBS Publishers and Distributors, Delhi.

References

Index

R

S

T

Symbol *f* and *t* refer Figure and Table respectively